WILD EDIBLE AND USEFUL
PLANTS OF IDAHO

WILD EDIBLE AND USEFUL PLANTS OF IDAHO

by Ray S. Vizgirdas

Illustrations by Edna Rey-Vizgirdas

RSVizgirdas
2017

To Mom and Dad...

Contents

Disclaimer

While this book documents the uses of wild plants found in Idaho, the author/publisher disclaims any liability for injury that may result from following any instructions for collecting, preparing or consuming plants described in this guide. Efforts have been taken to assure the descriptions of plants represented are accurate representations of the family, genus, and species noted. It should be understood that growth conditions, improper identification, and varietal differences, as well as an individual's own sensitivity or allergic response can contribute to a hazard in sampling or using a plant. Furthermore, the reader is encouraged to seek the assistance from experienced botanists in identifying any of the plants discussed in this book.

Preface and Acknowledgements

This is a revision of my previous book "*Useful Plants of Idaho*" published by Idaho State University Press in 2003. In *Useful Plants of Idaho*, there was a greater focus on the mountains environments and forests. In this updated edition, I now include the entire State – mountains, sagebrush steppe, grasslands, and anthropogenic.

In writing *Useful Plant of Idaho* I was grateful to many people for their review and helpful suggestions. I'd like to acknowledge those individuals again as their input is still seen in these pages. As before, special thanks goes out to Dr. Wayne Owen, Dr. Roger Rosentreter, Dr. Karl Holte, Dr. Richard Scott, Mr. Phil Delphey, and Mr. Bob Kibler. I am especially thankful to Mr. William Harwood and Dr. Trent Stephens at Idaho State University for making the 2003 publication a reality.

Since the 2003 publication, a whole new group of people provided support in a variety of ways. Their comments, suggestions, and ideas are reflected in the following pages. Specifically, I would like to thank the following individuals for their unique contributions (direct and indirect) to this new edition: Angela Staup, Dr. Don Mansfield, Jennifer Miller, Dr. Kerry McClay, Clay Lee, Dr. Jim Smith, Dr. John Cossel, Dr. Lynn Kinter, Kathryn Beall, Kristin Lundstrom, Dr. Mark Plew, Jennie Rylee, and Carolyn Volk.

INTRODUCTION

"I learned that a man may use as simple a diet as the animals and yet retain health and strength. I have made a satisfactory dinner off a dish of purselane which I gathered and boiled ... yet men have come to such a pass that they frequently starve, not for want of necessities, but for want of luxuries."

-- Henry David Thoreau

Look around ... almost everything you see is plant related. Many of the clothes you wear, the home you live in, this book, and the food you eat are derived from plants. A great majority of the medicines that keep us well also come from plants. In fact, if it weren't for plants, life as we know it would probably cease to exist.

Ancient humans learned very quickly that they needed food to survive. Over the years and through trial and error, humans developed and maintained a repertoire of edible, useful, and medicinal plants. It is because of these experiments that we have information on the uses of many plants. Unfortunately, today most people are unaware of these riches and have become dependent on only a few "domesticated" species of plants. Those that seek out wild plants find a continual source of stimulation and enjoyment, particularly as it relates to the survival of one's spirit.

This book describes some of the many uses of plants found in Idaho. It is intended to be used by wildlife enthusiasts, city dwellers, hunters, backpackers, fishermen, prospectors, campers, vegetarians, and students of wilderness survival. The use of technical terminology has been minimized.

No attempt has been made to suggest recipes in the preparation of food plants. In the field, simple seasonings such as salt, butter, lemon juice, and salad dressing are often all that is needed to make the wild plant foods more pleasing to the taste.

Finally, information on the medicinal uses of plants in this book is not intended to take the place of the local pharmacy.

AREA COVERED BY THIS HANDBOOK

Although this book provides a reasonably comprehensive guide to the wild edible, medicinal, and useful plants found within the borders of Idaho. However, it has utility in the surrounding states of Montana, Wyoming, Nevada, Utah, Oregon, and Washington as well.

SCIENTIFIC AND COMMON NAMES

In the discussion of wild edible, medicinal, and useful plants, plant families are arranged alphabetically within each of the four major groups of higher plants (ferns and their allies, gymnosperms, dicots, and monocots). While this arrangement may seem awkward to professional botanists, it has been adopted with the realization that this arrangement of families (and genera within families, species within genera) will be more easily consulted by readers in non-botanical fields who may have occasion to use this book. Where appropriate, I also provide information regarding changes in plant taxonomy based on input from the Angiosperm Phylogeny Group (APG). The APG system of plant classification is the first version of a modern, mostly molecular-based, system of plant taxonomy. Initially published in 1998, it was replaced by the improved APG II in 2003, APG III in 2009, and APG IV in 2016.

The scientific and common names are given for each species. A general description of the plants stressing key features of the families, genera, and some species are provided. Plant nomenclature (scientific and common names) used in this book follows that of the U.S. Department of Agriculture PLANTS database.

Common names for plants can be misleading and do not always distinguish among the species. Additionally, a species known by a common name in one region may have another common name elsewhere, leading to further confusion. However, common names have been retained since they are generally of more interest, and more likely to be known, by the public. In the end, however, you are encouraged to learn to identify plants by both their scientific and common names.

RARE AND PROTECTED PLANTS

The Native Americans were dependent upon nature for most of their needs and had an extensive knowledge of which plants were edible or useful. Because

Plants and People

It is easy to forget the importance of plants in our lives. Where the land is wild and rugged, plants are still vital for such basic human needs as food, shelter, medicine, and clothing. Humans have been recording the history of plant uses since early times. Expeditions to uncharted lands usually included botanists. Until recently, doctors of Western medicine studied botany because plants were the only source of ingredients for the treatment of disease. Enjoying the beauty of plants is just one of many reasons why protecting native species benefits all of us.

of this dependence, they shared a strong conservation ethic based on the sanctity of life. Today, with increasing human population and our demands upon natural resources, many species are becoming rare due to habitat destruction, competition with non-native species, or other means.

Many Idaho ecosystems have evolved in the absence of human activities and have no ready response to some kinds of disturbance. For example, though appearing rugged, mountain environments are fragile, highly susceptible to disturbance, and at times, have a low ability to rebound and heal after damage. The degree to which this is true is variable, but the vulnerability of mountain environments to disturbance is well documented (e.g., Price 1981, Zwinger and Willard 1972). Furthermore, the flora and fauna of mountain ecosystems are composed of species that are well-adapted to cope with environmental extremes, low productivity, and fluctuations within the system. Because of climatic extremes, a brief growing season, lack of nutrients at higher elevations, low biological activity and productivity, their island-like character, steepness of slopes, and the basic conservatism of the dominant life forms all make for the rate of restoration to original conditions after disturbance in mountain environments rather slow (Price 1981). These general concepts could also be extended to other ecosystems such as the sagebrush steppe, riparian, and grasslands.

Therefore, the plants you encounter will vary in their ability to withstand harvesting. For example, collecting berries may not directly kill a plant, but may affect its ability to survive. Additionally, because digging up the roots of a plant will destroy it, you must select your specimens carefully. Finally, keep in mind that many species of wildlife rely on plants for their survival as well, whereas most campers, hikers, hunters, and nature enthusiasts may collect and use plants for pleasure, not necessity.

It is strongly recommended that you obtain a list of threatened, endangered or sensitive plant species before you start collecting. In Idaho, the Idaho Native Plant Society, Idaho Department of Fish and Game, U.S. Forest Service, Bureau of Land Management, U.S. Fish and Wildlife Service, and various universities can be very helpful. Additionally, the land management agencies may have specific policies in place regarding collecting plants in state and federal lands. By avoiding rare species and using common sense, you should be able to enjoy wild plants without appreciably affecting either their population or surroundings.

NUTRITION AND SEASONALITY OF PLANTS

Wild plants are a good source of vitamins and minerals. In fact, much of the medicinal value originally associated with many wild plants was due simply to their high vitamin and mineral content. Wild plants can also provide proteins, carbohydrates, fats, vitamins, and minerals that are needed for good mental and physical condition. Certain amino acids, however, can only be obtained from animal products (e.g., meats, milk, cheese, eggs). Table 2 summarizes some of the important vitamins and minerals that are needed for good health.

Some Important Vitamins and Minerals for Health

Vitamins

These are organic compounds that are necessary in small quantities to prevent disease and help regulate the body's biochemical processes. Prolonged excessive doses of vitamins A, D, and K can have toxic effects. In addition to the vitamins listed below, biotin, choline, folic acid, and pantothenic acid are also essential nutrients.

Vitamin A	Vitamin A is not found in plants, but rather is manufactured by animals from pigments called carotenes which are common in plants. Vitamin A is essential for night vision, and promotes healthy skin and mucous membranes. It is also important for bones and teeth, proper digestion, and the production of red and white blood cells. It is fat soluble and sensitive to oxygen.
Vitamin B1 (Thiamine)	Found in both plant and animal tissues thiamine is important for the body's production of energy through the breakdown of carbohydrates. It appears to be important for normal functioning of nervous system and is involved in the action of the heart. Vitamin B1 is water-soluble and sensitive to heat. Most plants contain trace amounts.
Vitamin B2 (Riboflavin)	Riboflavin usually occurs in same foods as vitamin B1. It is essential for cell growth and enzymatic reactions by which the body metabolizes proteins, fats, and

carbohydrates. Vitamin B2 is water-soluble and sensitive to light.

Vitamin B6
(Pyridoxine)

B6 is still a relatively little-known vitamin. It participates in many enzymatic reactions and is particularly important for brain and nervous system function. Vitamin B6 is water-soluble and sensitive to oxygen and ultraviolet light.

Vitamin B12
(Cyanocobalamin)

Little or no B12 is found in plants. Strict vegetarians sometimes suffer from pernicious anemia, a disease associated with B12 deficiency. Vitamin B12 is necessary for proper functioning of cells, especially in the nervous system, bone marrow, and gastrointestinal tract. It is involved in the metabolism of fats, proteins, and carbohydrates. B12 is water-soluble and sensitive to light, acids, and alkalis.

Vitamin C
(Ascorbic Acid)

Vitamin C occurs in almost all plants to some degree. Since our bodies cannot make or store Vitamin C, a continuous supply must be present in the food we eat. Body cells require Vitamin C for proper functioning, as does the formation of healthy collagen (basic protein of connective tissue), bones, teeth, cartilage, skin, and blood vessels. Vitamin C also promotes the body's effective use of other nutrients such as iron, B vitamins, vitamins A and E, calcium, and certain amino acids. By promoting the formation of healthy connective tissue, Vitamin C helps to heal wounds and burns. Stress, fever, and infection tend to increase the body's need for Vitamin C. A deficiency of Vitamin C is called scurvy. Vitamin C is water-soluble and is sensitive to air, heat, light, alkalis, and copperware.

Vitamin D

Vitamin D does not occur in plants. However, some plants contain compounds called sterols, which when irradiated with ultraviolet light make Vitamin D. Vitamin D is necessary for healthy bones and teeth, and proper assimilation of calcium and phosphorus, and in preventing ricketts. It is a fat-soluble vitamin that is not sensitive to heat, light or oxygen.

Vitamin E
(Tocopherol)

Vitamin E is found in both plant and animal tissue. It is an antioxidant, acting to protect red blood cells, Vitamin A, and unsaturated fatty acids from oxidation damage. It also helps maintain healthy membrane tissue. In laboratory experiments, it was found to be necessary for fertility in rats. Vitamin E is fat-soluble, and is sensitive to oxygen, alkali, and ultraviolet light.

Vitamin K

While Vitamin K occurs primarily in plants, it is also synthesized by intestinal bacteria found in the small intestine. It is necessary for the synthesis by the liver of the blood clotting enzyme prothrombin. Vitamin K is fat-soluble and is sensitive to light, oxygen, strong acids, and alcoholic alkalis.

Niacin
(Nicotinic Acid)

A vitamin of the B complex, niacin occurs in both plant and animal tissue in various forms. In the body, niacin from plants is changed to niacinamide for use. Niacin takes part in enzyme reactions involved in the production of body energy and tissue respiration. Pellagra is a niacin deficiency disease. Niacin is water-soluble and is not sensitive to heat, acids, or alkali.

Minerals

These are chemical elements necessary for proper functioning of the body. Most are obtained from the foods we eat. There are two groups of minerals: macrominerals and microminerals. Macrominerals are found in relatively large amounts in the body, whereas microminerals are found in smaller amounts. Following is a list of minerals known to be necessary in human nutrition. There are other minerals, but their functions are not clearly understood.

Macrominerals

Calcium

This is the most abundant mineral in the body. It occurs in plants, dairy products, and seafood. Calcium is necessary for healthy bones and teeth, for clotting of blood, for the functioning of nerve tissue and muscles (including the heart), for enzymatic processes, and for controlling movement of fluids through cell walls.

Chlorine   As a gas chlorine is poisonous, but in the form of chloride compounds, it is an essential mineral. It acts with sodium to maintain the balance between fluids inside and outside the cells. Gastric juices in the stomach contain hydrochloric acid, the production of which requires chloride. Table salt (NaCl) is our main source.

Magnesium   Found in both plant and animal tissue. Magnessium is essential as an enzyme activator and is probably involved in the formation and maintenance of body protein.

Phosphorus   Occurs in plant and animal tissue. Phosphorus takes part in the production of energy for the body, and is second only to calcium as a constituent of bones and teeth. Phosphorus is necessary for metabolic functions relating to the brain and nerves, as well as for muscle action and enzyme formation.

Potassium   Potassium is abundant in plant and animal tissue. It promotes certain enzyme reactions in the body, and acts with sodium to maintain normal pH levels and balance between fluids inside and outside the cells.

Sodium   This is a common mineral in plants and animals. It regulates the volume of body fluids and balanced with potassium, it helps maintain cell fluid equilibrium. It is also necessary for nerve and muscle functioning. The ideal amount can be obtained through a diet of vegetables such as dandelion greens, spinach, mustard greens, watercress, and carrots.

Sulfur   Sulfur supplies come from sulfur-containing amino acids and from the B vitamins thiamine and biotin. Main sources are dairy products, meats, legumes, nuts, and grains. Sulfur is involved in bone growth, blood clotting, and muscle metabolism. It also helps to counteract toxic substances in the body by combining with them to form harmless compounds.

Microminerals

Copper   Found in plant and animal tissue. Copper is essential (along with iron) for the formation of hemoglobin in red blood cells. Copper is also important for protein and enzyme formation, as well as for the nervous and reproductive systems, bones, hair, and pigmentation.

Iodine   The only dependable source of iodine is found in seafood and seaweeds. Other plants will contain iodine if grown on iodine rich soils. It is necessary for normal physical and mental growth and development, as well as for lactation and reproduction. An iodine deficiency is called goiter.

Iron   Occurs in plant and animal tissue. The body retains iron very well and only trace amounts are needed in diet. Iron is essential to form the oxygen-carrying hemoglobin in red blood cells and involved in muscle function.

Manganese   Plants are the best source for manganese. Trace amounts are necessary for healthy bones and for enzyme reactions involved in energy production.

Zinc   Zinc is found primarily in animals, but also occurs in plants growing on good soil. It is important for various enzyme reactions, the reproductive system, and for the manufacture of body protein.

Categories of Foods

There are nine categories of plant foods discussed in this book. They include root vegetables, green vegetables, fleshy fruits, seeds, nuts, and grains, flowers, and inner bark. Root vegetables (i.e., tubers, corms, bulbs, rhizomes, true roots) include plants such as wild onions, camas, spring beauty, glacier lily, bitterroot, balsamroot, and knotweed. Roots are the storage organs high in carbohydrates. The greatest amount of energy from roots is available at the end of the growing season. These carbohydrates come in a variety of forms and flavors, and are not always readily digestible by humans. One type of carbohydrate found in some roots is inulin, which becomes sweet after cooking due to its conversion to fructose. If the skin of a plant's root is consumed, it can provide minerals and a small amount of vitamins.

Green vegetables include leaves, stems, shoots, and buds. Examples are fireweed (shoot and stem), lambsquarters, nettles, and mustard leaves. Many green vegetables are most palatable and digestible when they are young. Green vegetables are high in moisture, and often contain carotene, vitamin C, folic acid, and various minerals (e.g., iron, calcium, magnesium).

Fleshy fruits include serviceberry, gooseberries, currants, huckleberries, wild plums, cherries, and rose hips. Fleshy fruits are a good source of ascorbic acid, and contain high amounts of other nutrients such as calcium, Vitamin A, and folic acid.

Seeds, nuts, and grains are a good source of protein, fat, carbohydrates, vitamins, and minerals. Oils can also be rendered from these foods. Nuts are especially good sources of B vitamins, amino acids, and iron.

The cambium or inner bark of coniferous and deciduous trees and shrubs is another category of plant foods. The inner bark may be scraped off trees in the spring. Many species have high sap content. For example, maple sap is high in carbohydrate/sugar energy value for an inner bark food.

The final category of plant foods are the flowers. Rose petals, fireweed flowers, and mariposa lily buds are high in moisture. Flowers are low in proteins and fats, but some are rich in vitamin A (carotene) or vitamin C. There is little published information on the mineral content of flowers.

The nutritional value of plants changes with the seasons. During spring and summer, many plants are tender and rich in vitamins. Roots and tubers are high in carbohydrates and other nutrients. But as summer progresses, roots become less desirable because the stored energy is shifted to the aboveground parts. Fall is a time of nuts and berries, which provide a good source of protein. Roots again begin to store carbohydrates. Winter, however, can be bleak. The aboveground edibles may be limited to berries that have persisted into winter, bark and pine needles for teas, and inner bark. Teas can be restorative and do provide some food value. Teas can be upgraded into stews by adding insect larvae, birds, or mammals to make them more nutritious and sustaining.

Carbohydrates, Fats, and Proteins

The body's primary source of energy is glucose, a carbohydrate. Fats produce more energy for each gram consumed than carbohydrates. Fats and carbohydrates are the best sources of energy. Protein is the least preferred source of energy since it has to be extensively metabolized by the body to make glucose.

The nutritional value of plants also depends on preparation methods. For example, cooking greens in two changes of water makes them more palatable but can reduce the nutritional value. Generally, the preferred order of preparation for plants foods is: raw, quick-cook or steamed, baked, then boiled. Frying is the least desirable cooking method since it destroys many useful vitamins and minerals.

GUIDELINES FOR GATHERING

Since there are no general rules for distinguishing an edible plant from an unsavory or poisonous species, one must identify a plant correctly before attempting to use it. Some books even suggest that if you don't know a plant, you can eat a small quantity and wait to see if it has any adverse effects. This is a potentially serious mistake. For instance, if the unknown plant happens to be death camas (*Zigadenus*), not only would it cause much discomfort (such as a burning sensation in the mouth), it could kill you. Anyone who plans to search out and consume edible plants should exercise extreme caution. Correct identification of plants is necessary to avoid similar species or parts that may be unpalatable or poisonous. One of the best ways to learn about plants is to consult a knowledgeable botanist or qualified individual.

It is important to harvest plants with wisdom and respect. The uncontrolled harvesting of plants could severely damage delicate plant communities. In addition, it may be illegal to injure or uproot a living plant in some areas covered by this handbook (e.g., national parks and monuments). If a plant is rare or endangered, look for other edibles. If you are not in a survival situation, you should be even more frugal and considerate.

Also, be mindful of your own safety when gathering edible and useful plants. In Idaho and elsewhere, city, county, state, and federal agencies and organizations often spray chemicals to control noxious weeds, especially in areas where logging, mining, and grazing activities occur, or in developed campgrounds. While such chemicals may be considered to be "safe" - there are no guarantees. You should avoid collecting in areas affected by pollutants such as along roads or in drainages affected by mining activities.

In his book "*Stalking the Wild Asparagus*," Euell Gibbons described the taste of wild edible plants as an acquired one - ranging from awful to barely palatable. While we have sampled many wild plants that fall into those two categories, we've also found wonderful delicacies that make supermarket plant foods seem pale by comparison. If you take a positive outlook in your endeavor, it may someday help you if you were ever in a survival situation. Otherwise, it's a wonderful excuse to explore the western mountains.

A final thought:

"Where you find a people who believe that man and nature are indivisible, and that survival and health are contingent upon an understanding of nature and her processes, these societies will be very different from ours, as will their towns, cities and landscapes."

- Ian McHarg
Design with Nature (1969)

Section I - FERNS AND FERN ALLIES

Fern and fern allies include the club-mosses, horsetails, and ferns. They are herbaceous plants that reproduce by spores, which develop inside sporangia.

Water Fern Family - AZOLLACEAE

These are small floating plants on water.

Carolina Mosquito Fern (*Azolla caroliniana*)

General Description This is an aquatic fern that is triangular or polygonal in shape. It floats on the surface of the ponds, ditches, and other slow or sluggish waters. From a distance, it looks like a green-reddish carpet floating over the surface.

Field Notes/Uses Living within the cavities of this aquatic fern is a blue-green alga (*Anabaena azolla*). This nitrogen fixing alga excretes nitrogenous compounds into the cavity from which *Azolla* can absorb them, making nitrogen available for both species. In addition to fixing nitrogen by utilizing light energy, the blue-green alga associated with *Azolla* also releases hydrogen from water. This is the first reported known photosynthetic system for producing hydrogen from water that is stable in air and requires only water as a hydrogen source. In nature the fixed nitrogen combines with the hydrogen to form ammonia that fertilizes the host *Azolla* plant. However, in the laboratory, the fern-alga relationship can be diverted from producing ammonia to producing only hydrogen gas. While these are small-scale experiments to date, there may be some promise in producing hydrogen on a larger scale via this biological method. Hydrogen yields more energy (on a weight basis) than any other non-nuclear fuel, which is as important finding in light of dwindling petroleum supplies.

Azolla has other important contributions to many developing countries. In areas such as southeast Asia, Vietnam, and Singapore, it is used as a green manure. In these areas where rice is grown for human consumption, *Azolla* is deliberately cultivated. It is reported that in these rice fields, the yield of rice is 50 percent higher than normal. Pigs, cattle, and ducks are also fed *Azolla* in many parts of the world. In other areas, the plant is also useful in controlling mosquitoes and other weeds by blocking the water surface.

Horsetail Family - EQUISETACEAE

In this family, there is only one genus with about 23 species in worldwide. None of the species are of economic importance.

Horsetail (*Equisetum*)

General Description Thirteen 13 species of *Equisetum* occur in Idaho. In general, they are rhizomatous herbs with hollow, grooved, regularly jointed stems that are impregnated with silica. The leaves are reduced in size, appearing as a series of teeth around a joint. The spores are produced in cone-like structures atop the stems. They are found in moist soil along streams and rivers, marshes, and other damp habitats. The name comes from the Latin *equus*, for horse, and seta for bristles).

Field Notes/Uses Although all species are useful and identical in their application, it is *E. arvense* (field horsetail) most often discussed (Tilford 1993). The tough outer tissue can be peeled away and sweet inner pulp of all species can be eaten in small amounts. In large quantities, defined as greater than 20% of body weight by some authorities, they can be toxic (Willard 1992, Moore 1979). Certain chemicals in this plant are said to destroy specific B vitamins such as thiamine (Willard 1992). The enzyme thiaminase is apparently responsible for the poisoning. Cooking destroys this enzyme and renders the plants safe for consumption. The tuberous growth on the roots (actually rhizomes) can be eaten raw in the early spring or boiled later in the season.

Horsetails have an unusual chemistry. Some species contain alkaloids (including nicotine) and various minerals, whereas other species have been known to concentrate gold in their tissues, although not in sufficient amounts to warrant extraction. In the fall, the stems become impregnated with silicon dioxide and can be used to scour pots and pans or as a type of sandpaper for wood. Many Native Americans used horsetails to polish arrow shafts. The silica rich stems are reputed to be superior to the finest grades of steel wool. Additionally, the high silica content of this herb is said to be effective in strengthening bones and connective tissue (Tilford 1993).

Quillwort Family - ISOETACEAE

There are two genera and 75 species worldwide. None are of economic importance.

Quillwort (*Isoetes*)

General Description The common name is derived from the plants' resemblance to a bunch of quills. The approximately 75 species of quillwort worldwide can only be distinguished by microscopic examination of their spores. At least five species occur in Idaho. The spores are produced in sporangia (chambers) located in the leaf axils.

Field Notes/Uses None of the species is known to be of economic importance. It is reported they have been occasionally used as food by people in Europe (Pheiffer 1922). We found no record of anything injurious about the plants. The corm serves as a storehouse of food for the plant, as do the subterranean winter organs of many perennials; thus some food value is quite probable (Frye 1934). Weedon (1996) states that quillworts are rich in starch and oils and could be edible raw or cooked. The corms are edible, but not palatable.

Clubmoss Family - LYCOPODIACEAE

Clubmosses are low, evergreen, moss-like plants, often with prostrate stems trailing over the ground or spreading by rhizomes. The reproductive organs (sporangia) are produced in club-shaped spikes (strobili), hence the common name "clubmoss." In Idaho, the family is represented by three genera – *Huperzia*, *Lycopodiella*, and *Lycopodium*. The first two genera have only one species each - western clubmoss (*Huperzia occidentalis*) and inundated clubmoss (*Lycopodiella inundata*), both are found in northern Idaho. The genus *Lycopodium* is represented by at least 7 species throughout most of the state.

Clubmoss (*Lycopodium*)

General Description Approximately 450 species of clubmosses occur worldwide, with seven species found in Idaho. They have small, narrow, evergreen leaves that are alternate and more or less spirally arranged. The spores are produced in spore cases (sporangia) and are found in the leaf axils.

Field Notes/Uses The spores of clubmoss are the primary product of importance for this family. In fact, in Germany clubmoss spores were used as early as 1665. The fine, yellow spores are very light, odorless and tasteless. The spores do not contain any medicinal substances, but were used by aboriginal peoples as a drying agent for wounds, a treatment for nose bleeds, and as a dusting powder to prevent chaffing. The spores' fine texture makes them smooth and prevents stickiness. The spores of *L. clavatum* (running clubmoss) were used in decoctions to increase urine flow, to treat severe diarrhea, and to increase one's appetite.

To collect spores, the tops of the plants were dried in vessels to prevent the spores from blowing away. After drying the plants were beaten and rubbed, and the spores sifted out of the vegetation.

The spores of all clubmosses are also very flammable and were used in fireworks, stage lighting, and early flash photography (Pojar and MacKinnon 1994). When heated slowly, the spores burn quietly, but thrown into a flame, the spores ignite explosively and produce a lot of light. The spores also contain a waxy substance which makes them water repellant. It is said that if you coat your hand with clubmoss powder and submerge it in water, your hand will remain dry.

Warning Most clubmoss species contain poisonous alkaloids, and therefore, should not be taken internally.

Waterclover Family - MARSILEACEAE

The family is comprised of three genera and 65 species distributed worldwide. Two genera are native to the U.S. They are of limited importance as ornamentals.

Hairy Pepperwort (*Marsilea vestita*)

General Description While there is about 65 species of waterclover known to occur worldwide, two species can be found in Idaho. Of the two, it is *M. vestita* (hairy pepperwort) that is more widely ranging. It is found in shallow lakes and ponds or on their muddy borders. It has also been found along slow moving streams. Arising from the rhizome are the distinctive leaves arranged like those of a four leaf clover. The leaflets are pale green and hairy. Spores are produced within an elliptical, nut-like body borne on short stalks.

Field Notes/Uses While there are no documented uses for this species in Idaho, related species are supposedly edible when dried and ground into flour (Weedon 1996). The sporocarps of Nardoo (*M. drummondii*) found in Central Australia can be crushed into a powder between stones, made into a dough by adding water, and baked into cakes (Frye 1934).

Adder's-tongue Family - OPHIOGLOSSACEAE

There are four genera and 70 species of this family. Two genera are found in the United States. None of the members of this family are of economic importance.

Grapefern (*Botrychium*)

General Description Worldwide, there are about 40 species of grapefern, of which 14 occur in Idaho. This is a difficult genus to study and requires fully developed specimens for accurate identification. The spore sacs, technically termed the sporangia, are borne in grape-like clusters on a naked stalk, not on leaves as in the true ferns.

Field Notes/Uses One species, *B. lunaria* (common moonwort) is found in grassy meadows and edges of woods or in open woods at various elevations. Early records appear to indicate that juice extracts from this species were used to stop bleeding and vomiting, and also for the treatment of bruises. They may have been used to concoct balsams for healing internal wounds (Grillos 1966). Foster and Duke (1990) describe a root poultice or lotion made from *B. virginianum* (rattlesnake fern) that was used for snakebites, bruises, cuts, or sores. In folk medicine, the root tea was used an emetic.

"FERN FAMILY (Polypodiaceae)"

In actuality, the various species of ferns in Idaho are found in different families. They include Dennstaediaceae, Dryopteridaceae, Polypodiaceae, Pteridaceae, and others. For convenience, they are treated here under the "fern family" heading, but their current taxonomic position is also highlighted. Generally speaking, fern fronds can be used as makeshift place mats and interspersed between layers of food in cooking pits.

Caution A number of survival handbooks suggest that most ferns in their young stage are edible. However, in most cases, they are referring to bracken fern (*Pteridium*). It is best to avoid ferns unless you are able to positively identify them. Many plants such as poison hemlock (*Conium maculatum*) have delicately cut leaves, which to the untrained eye could easily be mistaken for a fern.

Spleenwort Family - Aspleniaceae

Sleenwort (*Asplenium*)

General Description This is a genus with about 650 to 750 species of terrestrial and epiphytic ferns with divided leaves. These are ferns that grow in moist crevices of cliffs and ledges or in soil, usually in cool shaded canyons. The sori are elongate and occur on the veins.

Field Notes/Uses The genus name comes from the Greek *a* (meaning not), and *splen* (meaning spleen), alluding to the plants supposed medicinal properties. For example, the Cherokee made an infusion of the plant for "breast diseases" and for coughs. Several species are grown as ornamentals.

Chain Fern Family - Blechnaceae

Deer Fern (*Blechnum spicant*)

General Description Deer fern fronds are dimorphic: the sterile leaves are evergreen and are spreading or appressed to the ground; the fertile leaves are fewer in number, deciduous, and much longer than the sterile leaves. Deer fern has woody rhizomes

Steam Pit Cooking

The steam pit cooking method has been used by many different peoples around the world for thousands of years. This method of cooking locks in the natural juices and flavor of the food during the cooking process.

There are many variations of the pit, but basically all you need to do is simply dig a hole in the ground about two feet deep and two feet across, and line the bottom and sides with flat rocks. One rule of thumb is to have the hole no more than three times the size of the total amount of food. Construct a fire in the pit until the rocks are red-hot. After about an hour or so, carefully remove the coals and place about 6-8 inches of green grass, fern fronds, or other non-poisonous vegetative matter on top of the hot rocks. Then place your food on top of this layer of plants and cover with an additional 6-8 inches of grass. On top of the grass, place a thin layer of bark slabs or brush to prevent dirt from sifting through to the food. Finally, cover with a mound of dirt and leave it alone for a few hours. This will give you time to go do other things.

Field Notes/Uses Deer fern has a sporadic circumpolar distribution. In North America it is distributed from coastal Alaska to California. Deer fern occurs mostly west of the Cascade Range but is also found in northern Idaho.

For some Native Americans this fern was considered a starvation food. The young tender stalks were peeled and the center portion was eaten. It was also considered a spice plant. In this case the fronds were used for flavor in cooking

Bracken Fern Family - Dennstaedtiaceae

Western Brackenfern (*Pteridium aquilinum*)

General Description This is a widely distributed species found in open woods, rock slides or slopes in damp or dry places, up into the high mountains. The generic name comes from the Greek *pteris*, a wing, and is applied to ferns because of their feathery leaves.

Field Notes/Uses The young fronds, or fiddleheads of bracken fern can be collected, boiled and dried in the sun. The dried product can then be used as a winter food. Old fronds may be poisonous in large amounts. The starchy rhizome (underground stem) is edible after roasting or boiling, but is usually tough. The leaves can be used as one of the protective plant layers for pit cooking. Some Native Americans would consume only fiddleheads so that their scent would not scare off deer. A root tea was used for stomach cramps and diarrhea, and poulticed roots were used for burns and sores (Foster and Duke 1990). Ashes of the plants have been used as an ingredient to make glass and soap.

Caution While this plant has traditionally been accepted and harvested as a suitable edible, there is new evidence indicates that eating sufficient quantities over a period of time may be dangerous to your health (Foster and Duke 1990, LST 1972). The plant is known to contain several poisonous compounds, including a cyanide-producing glycoside (prunasin), an enzyme, thiaminase, that reduces the body's thiamine reserves, and at least two potent carcinogens, quercetin and kaempferol. Another, unidentified toxin is believed to be naturally occurring, radiation-mimicking substance, also apparently mutagenic and carcinogenic. Bracken has caused many livestock deaths. The risks to humans of eating bracken fiddleheads and rhizomes have not been fully established, but thier safety is questionable (Kuhnlein and Turner 1991). Schofield (1989) reports that it is currently suspected of causing stomach cancer in Japan.

Wood Fern Family - Dryopteridaceae

Ladyfern (*Athyrium*)

General Description The two species of ladyfern in Idaho are medium-sized plants growing in moist, shady places. The leaves are 1-3 pinnate and clustered. The name is from the Greek a, meaning without, and thurium, referring to a long oblong shield. One species, *A. filix-femina* (common ladyfern) is widely distributed in North and South America

Field Notes/Uses While edibility of the two species that occur in Idaho are unknown, Pojar and MacKinnon (1994) indicate that *A. filix-femina* fiddleheads were eaten in the early spring when they were 2 to 6 inches tall. They were boiled, baked or eaten raw with grease. Grillos (1967) further indicates that common ladyfern has certain chemical properties for medicinal use. The underground stems, pulverized to a powder have been used to drive worms out of the intestinal system, although this use has not been medically recognized. A tea made from the root was used as a diuretic, and the powdered root was used externally for sores (Foster and Duke 1990

Bladderfern (*Cystopteris fragilis*)

General Description This 6 to 16 inch tall fern is loosely tufted from a short creeping rhizome. The leaves are thin and delicate in texture. The stipes are brown below, yellowish above, and smooth. The indusia are small, attached at one side and arching back to form a hood.

Field Notes/Uses This is a widely distributed fern, and is found in the crevices of cliffs and ledges, in soil under rocks, shrubs, or trees. The genus has been placed in the family Cliff Fern family (Woodsiaceae).

The plant was used as a dermatological aid by the Navajo. Here, a cold, compound infusion of the plant was made and used as a lotion for injuries.

Woodfern (*Dryopteris*)

General Description Four species of woodfern occur in Idaho. The name is from the Greek *drys*, meaning oak, and *pteris*, meaning fern.

Field Notes/Uses The edibility of the various species of woodfern is unknown, but some species are reported to be edible, poisonous, or of medicinal value. For example, the rhizome of *D. expansa* (spreading woodfern) was used as food after baking them in a pit overnight. The fiddleheads, with the chaffy coverings removed, were boiled and eaten with seal oil and dried fish by Alaska

Eskimos. The rhizomes of *D. filix-mas* (male fern) were also eaten raw or steamed.

Medicinally, several species contain phloroglucinol derivatives ("filicin"), which paralyze intestinal parasites. For example, a root tea from *D. cristata* (crested woodfern) was traditionally used to induce sweating, clear chest congestion, and expel intestinal worms (Foster and Duke 1990). In addition, an oleoresin was extracted from the roots of *D. filix-mas* (male fern) to expel worms.

Western Oakfern (*Gymnocarpium dryopteris*)

General Description This is one of three species in Idaho. It is a small and delicate fern. The deciduous leaf blades are light green and the petioles are 4-12 inches long and are oriented parallel to the ground. The glabrous to slightly glandular blade is broadly triangular in outline and distinctively divided into 3 triangular leaflets that are 1-2 times pinnately divided into paired asymmetric pinnae. The basal pair is evidently the largest.

Field Notes/Uses Oakfern occurs in moist to wet forests, as well as in some riparian areas. This species does well in shade so deep that other undergrowth species cannot exist. *Gymnocarpium* means "naked fruit," because the spore clusters are not covered by an indusium.

Oakfern is a characteristic indicator species of moist sites. As moisture levels increase other species such as horsetails (*Equisetum*) become more abundant. The Cree crushed the leaves to repel mosquitoes and soothe mosquito bites.

Swordfern (*Polystichum*)

General Description Swordferns are usually observed on rocky materials, particularly in damp, wooded places. The name is from the Greek *polys* (many), and *stichos* (row), because the sori of some species develop in several rows. Seven species of *Polystichum* can be found in Idaho.

Field Notes/Uses In general, the leaves of these ferns can be used as a protective layer in pit cooking, as flooring or bedding. The large rhizomes of *P. munitum* (swordfern) can be roasted over a fire or steamed in a pit oven, then peeled and eaten. The cooked rhizomes are also said to be a cure for diarrhea (Pojar and MacKinnon 1994).

Woodsia (*Woodsia*)

General Description These small ferns commonly grow in rocky places. The underground stem is densely tufted and clothed with broad, thin scales. The leaves are clustered, numerous, small, linear to lanceolate-ovate, and once- or twice-pinnate. The sori are round and seated on the back of the free veins, and the indusia is under the sori with star-shaped divisions.

Field Notes/Uses The genus name honors Joseph Woods, an English botanist. Rocky Mountain woodsia was used as a sign of water when traveling through the mountains by the Natives further to the north. Some species are cultivated as ornamentals.

Western Polypody (*Polypodium hesperium*)

General Description This is a shade-loving fern. The fronds of are pinnatifid to nearly pinnately compound and attached to creeping rhizomes by a distinct articulation. The sori are round in outline, and there is no indusium. This is one of two species in Idaho growing on rock ledges and crevices. The genus name is Greek, meaning many and foot, alluding to the numerous protuberances on the rhizomes.

Field Notes/Uses The rhizomes were chewed or an infusion of the rhizomes was made for colds and sore throats. The rhizomes were also used as medicine for sore gums. The rhizome has a pleasant, sweet taste, almost like licorice. The roots of the other Idaho species - licorice fern (*P. glycyrrhiza*) - were also eaten by Native Americans. Other Natives Americans used bruised roots to heal sores and to relieve rheumatic pain.

Maidenhair Fern Family – Pteridaceae

Northern Maidenhair (*Adiantum aleuticum*)

General Description Tw hundred species of *Adiantum* occur worldwide. This is a widespread species that is found in the low to middle elevations on up to the subalpine. Look for it in areas with rich moist soils near streams, shady forests, and in the spray zones of waterfalls.

Field Notes/Uses This black-stemmed fern can be used in basketry. The herbage is reported to be bitter and causes an increased secretion of mucus. The leaves were used as a tea or syrup to treat colds, coughs, and hoarseness. Rhizomes were used as a stimulant, to soothe the mucous membranes of the throat and loosen phlegm (Willard 1992, Foster and Duke

1990, Moore 1979). Some of these medicinal uses have been recorded since Classical times.

Cliff-brake (*Pellaea*)

General Description These are small tufted ferns growing in crevices of rocks. The rhizomes are short, thick, creeping and densely brown scaly, and covered with old stipes. The leaves are singly pinnate or bi-pinnate. The stipes are dark reddish-brown in color. The sori are covered by the recurved margin of the leaf segments. The generic name comes from the Greek *pellos* (dusky), and refers to the appearance of the leafstalks.

Field Notes/Uses A refreshing tea can be made from a related species in the southwest (*P. mucronata*) by Native Americans. They also used the tea medicinally as a decoction to stop hemorrhages, to reduce fevers, and as an emetic. The decoction was used as a wash for skin problems. Other Native Americans used the brown fibers from the rhizome to make basketry patterns.

Caution The younger leaves and stems are occasionally eaten by sheep and other grazing animals. However, they are poisonous and frequently cause death when eaten.

Goldback Fern (*Pentagramma triangularis*)

General Description This perennial fern has a rhizome with slender scales that have a thickened, blackish midstrip and thinner, brown, narrow margins; petioles chestnut-brown to purplish-brown, which is notable longer than the blades. The blades are glabrous above and have a pale to bright yellow waxy powder on the underside. The few pinnae or segments are opposite, and the lowest pair is the largest. The leaf margins are very narrowly revolute, but not covering the sporangia. Goldback fern is common in moist habitats.

Field Notes/Uses Native Americans in California and Nevada chewed the plant for toothaches. To other Native Americans, the plant was used to mitigate the afterpains of childbirth

Spikemoss Family - SELAGINELLACEAE

This is a family of moss-like plants that includes many important natural ground covers on rocky or gravelly soils. Members are sometimes called "little clubmoss" because of their resemblance to the Clubmoss Family (Lycopodiaceae).

Little Clubmoss (*Selaginella*)

General Description Five species of little clubmoss can be found in Idaho. They can be found in a variety of habitats including dry rocks, ledges, cliffs, crevices, talus slopes, meadows, and lake shores up to timberline.

Field Notes/Uses The spores of little clubmoss were once used like that of clubmoss (*Lycopodium*) in making pills not adhere to one another. Many campers still gather the plants with mosses and other plants to make a soft bed when sleeping on the ground.

While there are no edible or medicinal uses recorded for Idaho species, related species, such as *S. concinna* and *S. obtusa* found on the Reunion Islands in the Indian Ocean were used medicinally as astringents, blood purifiers, and as carminatives in cases of dysentery. In Mexico, a variety of *S. rupestris* was used as a home remedy, and in the East Indies, *S. convoluta* was considered to be an aphrodisiac (Frye 1934) .

Section II - GYMNOSPERMS

Cypress Family - CUPRESSACEAE

There are 19 genera and 130 species in this family. Five of the genera are native to the United States. In Idaho, only *Juniperus* and *Thuja* occur. The family is economically important as a source of timber trees and ornamentals.

Juniper (*Juniperus*)

General Description Junipers are evergreen shrubs and trees that usually grow in dry, rocky soils. The leaves are scalelike or awl-like. In some cases, both types of leaves may be found on the same plant. Four species of *Juniperus* may be encountered in Idaho; *J. communis* (common juniper), *J. scopulorum* (Rocky Mountain juniper), *J. osteosperma* (Utah juniper), and *J. occidentalis* (western juniper). These plants are often referred to as cedars because of the fragrance given off when the wood is burned.

Field Notes/Uses Common juniper is a low spreading shrub, approximately 2-3 feet high. It differs from the other species in that the needles are awl-shaped, green, with a white line on the upper surface. The fleshy cones (often referred to as berries) are bluish in color. This species occurs in dry forests and open slopes from the foothills to above timberline. The Rocky Mountain, Utah, and western junipers are more tree-like with evergreen, scale-like leaves, often with a grayish-green to silvery tinge. The cones are light blue, usually with 2-3 seeds. They are found at lower elevations than the common juniper.

> ### Ash Cakes
>
> *Make dough by mixing flour with water. Pat the dough into a patty. The thicker it is the more doughy it will be, whereas the thinner it is the crispier it will be. Throw the patty into the ashes (hot coals), let it cook a bit, and you have an ash cake.*

Junipers offer countless products. The "berries" (actually fleshy cones) and twigs can be used to make tea, season game, smoke fish, repel moths, soothe rheumatic pains, and kill infectious germs. The fleshy cones are edible raw, but taste better if dried, ground, and used as a flour, flour extender, or made into cakes. Cooking the flour with other foods can make it more palatable. The berries can also be roasted and ground, and used as a coffee substitute. The Swedes make an extract from the berries, which they generally eat with bread, in much the same way we use butter. In addition, the berries have been used to give gin its characteristic flavor.

The boughs can be steeped in hot water for 5-10 minutes to make a beverage. Cooking them in an uncovered pot is recommended to allow the volatile oils to escape. A leaf or berry infusion was used to relieve urinary problems.

The shredded bark is excellent tinder for primitive fire starting techniques and can be used as bedding and padding. It is said that some Native American tribes ate the inner bark in times of famine (Strike 1994). The inner bark was also used for clothes and mattresses and could be worked with the hands until soft enough to use for baby diapers or sanitary pads.

Juniper oil extract can be rubbed on skin as an insect repellent (Tull 1987), and to relieve pain in muscles and joints (Tilford 1993). The bark, roots, twigs, and cones furnish red dyes. The white film covering the berries is a type of yeast that can be used to make a primitive sourdough starter.

Juniper wood was used for bows by many Native Americans. In many cases, it was considered the best wood for sinew-backed bows, even though it was difficult to work with. The wood also makes good ladles and other implements.

Caution Junipers are known to have diuretic properties. Juniper tea should be used only in moderation and should not be consumed by pregnant women since it may cause uterine contractions (Tyler 1987).

Nature's Yeast

The white dust covering the bark of aspen trees, Oregon-grape berries, and Juniper berries contains within it a type of yeast. It is possible to utilize this natural yeast as leavening, to give those ash-cakes more body. Start by collecting some of the berries or bark, add a cup or so of flour, and warm water to the mixture in a container with a loose fitting lid. Let it stand for a day or more.

To use the starter you've created, mix it with flour at a ratio of about 1 to 5 (i.e., 1 cup starter to 5 cups flour). Add enough water to make it manageable, and knead. In the outdoors setting, place the dough on a warm rock and allow it rise. In a couple of hours, it will probably only rise slightly. After rising, place the sourdough in a rock oven and bake and be careful not to burn the loaves.

The starter described here often does work as well as commercially available yeast. Under primitive conditions experimentation is the rule. Remember to replenish the starter each time you use it by adding water and flour. The more you use the starter, the better it becomes.

Western Redcedar (*Thuja plicata*)

General Description This is the only member of the genus in Idaho. The wood and leaves have a pleasant fragrance. *Thuja* is Greek for an evergreen tree. Western redcedar wood is useful and valuable. The heartwood resists decay and is used in shingles, shakes, utility poles, fence posts, and siding for homes. Western redcedar is usually found at middle elevations in northern Idaho.

Field Notes/Uses Leaves of a related species, T. occidentalis (white cedar), were used as a tea by Ojibwa Indians in Canada and possibly other tribes in eastern North America (Kuhnlein and Turner 1991). Kuhnlein and Turner (1991) also indicate that Blackfoot Indians may have eaten the inner bark of T. plicata fresh or pressed into cakes. This use is not widely reported, however.

Willard (1992) indicates that redcedar was used occasionally for medicinal purposes. Nez Perce Indians made a bitter tea of the boughs to relieve coughs and colds, and chewed the buds to ease toothaches.

Western redcedar wood was used to make important cultural items such as dugout canoes, harpoon shafts, boxes, posts, dishes, arrow shafts, fish weirs, drum logs, and paddles. The wood can be used to make a drill and hearth for starting fires by friction, and the wood burns with little smoke. The fine roots or shredded bark can be used to make baskets and containers.

Warning Redcedars (*Thuja*) contain thujone, a substance that if taken in high concentrations may cause convulsions and other disturbances.

Pine Family - PINACEAE

There are 10 genera and 250 species in this family worldwide. Six genera are native to the United States and Idaho. The pine family is comprised of trees with needle-like leaves. Except for larches (*Larix*), most species are evergreen. Many are economically important timber trees.

True Firs (*Abies*)

General Description Worldwide, there are about 50 species of *Abies*, three of which occur in Idaho. They are trees of cold climates with short, flat, blunt needles that grow directly from the branch. The needles are grooved above, with white stripes beneath and sometimes above. Cones mature in late summer and fall apart, leaving behind a stalk. The wood is very light, soft, and brittle, and has been used to manufacture boxes and crates. Firs are also important as pulpwood, and they make good Christmas trees since they retain their needles even when dry. The young bark often produces conspicuous pitch blisters, containing strong-smelling liquid oleoresins.

Field Notes/Uses Subalpine fir (*A. lasiocarpa*) is a widespread tree at high elevations. The inner bark can be eaten raw. Medicinally, the resin, or sap was brewed as an emetic to "clean out the insides" (Willard 1992). Leaves were used in a poultice for fever and chest colds (Hart 1996). The sap was also chewed for pleasure and as a cure for bad breath. Additionally, Hart (1996) describes its use by many Native Americans for cuts, wounds, ulcers, sores, skin infections, colds, coughs, and constipation. The needles were pounded into a fine powder, mixed with grease and then rubbed on the infected skin. The sappy or gummy secretions were also used as an antiseptic for wounds. The boughs can be used for bedding, covering floors, and as incense.

Grand fir (*A. grandis*) is commonly encountered and is also known as "stinking fir" because of an odor that may be unpleasant to some people. A brown dye can be made from the bark, and the needles were boiled to make a medicinal tea by some Native Americans (Coffey 1994, Moore 1979).

Larch (*Larix*)

General Description Larches are the only northern conifers that drop their needles during fall. There are about 12 species of *Larix* worldwide. Of the two species in Idaho, *L. occidentalis* (western larch) is much more common than *L. lyallii* (subalpine larch). Western larch is found in mid-elevation forests whereas subalpine larch is restricted to high mountaintops above 7,500 feet. Larch is said to have one of the heaviest, strongest, and toughest softwoods. Because of its durability, it is used for fences, posts, poles, and ties.

Field Notes/Uses The inner bark was reported to be eaten by some aboriginal peoples (Kuhnlein and Turner 1991). Sweet tasting syrup is derived from western larch was collected by the Flathead and Kootenai Indians. They obtained this syrup by hollowing out a cavity in the trunk, allowing about one gallon of the sap to accumulate. Natural evaporation then concentrated the syrup making it sweeter (Hart 1996). The syrup is a natural sugar called galactan that is found in the wood. Its taste resembles a slightly bitter honey. The syrup can also be made into medicine and baking powder.

The bark was made into poultices and used for chronic eczema, psoriasis, bruises, and wounds (Willard 1992). The strong root fibers were used for sewing and basket weaving. The solidified pitch (sap hardens when exposed to air) was also chewed as gum.

Spruce (*Picea*)

General Description Of the three species that occur in Idaho, the more widespread species is Englemann spruce (*P. englemannii*). It is a large tree up to 200 feet tall. The bark is thin, grayish, and scaly. The needles are quadrangular, stiff and sharp pointed, with white bands on all four sides. The cones develop in the tree tops and are about 1 to 3 inches long. Englemann spruce is found on cool, moist slopes up to the alpine zone usually with subalpine fir (*Abies lasiocarpa*). Blue spruce (*P. pungens*) which reaches its northern limits in eastern Idaho is more common in the Middle and Southern Rocky Mountains. The third species is white spruce (*P. glauca*) and it is found in extreme northern Idaho.

In general, spruces have a light, soft, compact, straight-grained wood that is stiff, strong, easy to work, and comparatively free of resin. Spruce are used for pulpwood, light construction, boxes, millwork, and Christmas trees. The resonant wood is also used for sounding boards for pianos and the bodies of violins and similar instruments.

Field Notes/Uses The cambium or inner bark of Englemann's spruce can be eaten raw, boiled like noodles, or dried and used as a flour substitute (Pojar and MacKinnon 1994). The inner bark was also used as a laxative. Some aboriginal peoples ate the new growth raw, a good source of Vitamin C (Brown 1983). The green spring growth on branches can be steeped in hot water for tea. Spruce beer is made from fermented needles and twigs that have been boiled with honey. The gum can be applied to cuts and wounds as a healing agent. It can also be applied to the skin to protect against sunburn (Willard 1992). The crushed needles can be rubbed on traps and skin to camouflage human scent. Although spruce seeds are very small, they are edible raw or cooked.

Pine (*Pinus*)

General Description Six species of *Pinus* occur in Idaho including western white pine (*P. monticola*), limber pine (*P. flexilis*), whitebark pine (*P. albicaulis*), ponderosa pine (*P. ponderosa*), singleleaf pinyon (*P. monophylla*), and lodgepole pine (*P. contorta*). Pines have needles in bundles of 1, 2, 3, or 5, with a membranous sheath called a bundle sheath at the base. Pines may be divided into two major subgroups - "soft" pines and "hard" pines. In soft pines, the needles are usually in bundles of five and the cones have no prickles.

Field Notes/Uses The wood of the soft pines is straight grained, that is comparatively free from resin, and easy to work. It is used for rough carpentry, cabinetwork, patterns, toys, crates, and boxes. In contrast, hard pines have 2-3 needles per bundle, and the cones have prickles. The strong, resinous wood of hard pines is used in buildings, bridges, ships, and other types of heavy construction. Because of its durability, the wood of hard pines is valuable for floors, stairs, planks, and beams.

All pines have edible seeds. However, they are an erratic food source, yielding an abundant crop in some years and a sparse crop in others. To collect the cones, long poles were used to knock them from the branches. One of the best ways to gather seeds is to heat the green cones until they open. The seeds are best when harvested in fall or early winter when cones normally release their seeds. The nutritious seeds can then be shelled and eaten, or ground or roasted and made into flour. Seeds may contain as much as 15% protein, 62% fat, and 18% carbohydrates, with approximately 3,000 calories per pound (Harrington 1967, Vizgirdas 2003).

The inner bark is also edible in an emergency. Though tedious, the tender mucilaginous layer between the bark and wood was scraped or peeled off. It was then cooked or ground into meal. The use of the inner bark by Native Americans was so extensive that early explorers reported recorded large areas of trees stripped of bark (Fernald and Kinsey 1958).

The firm and unexpanded pollen cones can be boiled and eaten. They have a surprisingly sweet and non-pitchy taste. The edible pollen is usually mixed with flour and used as a soup thickener (Brown 1983, Fernald and Kinsey 1958).

The needles of most pines can be steeped in hot water to make a satisfying tea and are a good source of Vitamin C. It also takes some practice to steep the right amount of leaves, since too much may be too strong. Additionally, the pine cleaning fluid can be extracted from boiling the needles and skimming off the oil-like substance from the surface. It may take a lot of pine needles to get a small cupful.

Pine Pitch Glue

Pine sap is very useful in the construction of a number of wilderness tools. To make glue or cement, first melt the sap and add finely ground charcoal. The resulting glue is somewhat hard and brittle. By experimenting with other materials such as sand or egg shells, you can develop glues with different properties. The pitch is also used for waterproofing coiled or tightly woven baskets. The pitch is applied inside and outside of the container.

Pine sap can be collected in quantity from cuts, burns, and broken branches. The collected sap is then heated and formed into balls for future use. Be careful not to expose the sap to flames as it is very flammable.

All species of pine have been used medicinally for many centuries. Chewing the sap was said to be soothing for a sore throat. The sap can be dried, powdered, and applied to the throat with a swab. It was also heated and used as a dressing to draw out embedded splinters or to bring boils to a head. Smeared on a hot cloth the sap was used much like a mustard plaster in treating pneumonia, sciatic pains, and general muscular soreness (Willard 1992, Brown 1983, Moore 1979). Pine oil is widely used in massage oil for treating muscular stiffness, sciatica, rheumatism, and in vapor rubs for bronchial congestion. All pines are rich in resin and camphoraceous volatile oils, including pinene, that are strongly antiseptic and stimulant. The needles and resin produce a brown dye.

Pine roots were valued as twining material for baskets. The roots, about as thick as a pencil, can be several feet long. Roots were collected from opposite sides of a tree each year to prevent destroying the tree. After cleaning, roots were buried in a heated pit. Fire built over the pit for two days turned the roots a light tan color and they were then ready to be split into smaller strips.

Douglas-fir (*Pseudotsuga menziesii*)

General Description This is the only species of the genus in Idaho. The needles are flat, soft, and flexible. The cones hang down, with protruding 3-pronged bracts between the scales. Douglas-fir is an important timber species. The wood is resinous with close, even, well-marked grains, and is of medium weight, strength, stiffness, and toughness. It is very durable, and when well-seasoned, does not warp. It is used in piles, ties, floors, and millwork, and to make a variety of items such as spear handles, spoons, fire tongs, and fishing hooks.

Field Notes/Uses A tea can be made from the needles of Douglas-fir. Similar to pines, the pitch can be used as glue. It can be used for sealing implements and caulking water containers. Medicinally, the sap provided a salve for wounds and skin irritations (Pojar and MacKinnon 1994). The pliable roots have been used in weaving.

Hemlock (*Tsuga*)

General Description Two species of *Tsuga* may be encountered in Idaho, mountain hemlock (*T. mertensiana*) and western hemlock (*T. heterophylla*). The needles are up to one inch long. The light brown, papery cones are about $1\frac{1}{2}$ inches long. The trees require the cool climate, high humidity, and plenty of moisture that is found in the northern half of Idaho.

Field Notes/Uses Hemlocks furnish a cheap coarse lumber for framing, sheathing, lathes, rafters, and other types of rough construction. Some other uses include pulp, ties, boxes, and planks. The wood is coarse grained and splintery, but is very strong, stiff, tough, and easily worked. Western hemlock is one of the best pulpwoods. It is also a source of alpha cellulose for making cellophane, rayon yarns, and plastics.

A tea can be made from the needles or inner bark of either species by steeping them in hot water. The inner bark can also be eaten raw, boiled like noodles, cooked with berries or dried and used as a flour substitute, or it can be blended with sap and pressed into cakes. Indians of southeastern Alaska made coarse bread from the inner bark. Both species are high in tannins and can be used as a tanning agent, pigment, and cleansing solution. A red dye was derived from the bark to color basket materials. The wood is durable and easy to carve into implements such as spoons, combs, spear shafts, and fishing hooks. The branches make a good bedding material. Hemlock pitch is used topically as a poultice, liniment, and salve. The young branch tips were chewed by aboriginal peoples as a hunger suppressant (Pojar and MacKinnon 1994). Finally, the powdered inner bark is a good ingredient in foot powder, as it reportedly eliminates foot odor.

Yew Family - TAXACEAE

There are five genera and 20 species distributed in the northern hemisphere. Two of the genera are native to the United States. The family is of economic importance as a source of timber and ornamentals.

Pacific Yew (*Taxus brevifolia*)

General Description This is an understory shrub in coniferous forests, but in favorable habitats can be a tree. In Idaho, the greatest density of this species occurs in the Nez Perce National Forest.

Field Notes/Uses Kingsbury (1964) describes yews, in general, as toxic. However, the twigs of some eastern species were used for tea. Additionally, the red fruits borne on the female plant are said to be edible in small amounts and have a cherry jello-like taste, but the seeds are considered to be very poisonous.

The bark of Pacific yew contains taxol, an anti-carcinogen that slows tumor growth. It is presently being tested against breast, ovarian, and kidney cancer. This species has become an important source of income in the Northwest, but because of its slow growth, there is growing concern that the Pacific yew may become endangered due to overharvesting.

The leaves are poisonous if eaten. However, some Native Canadians used the leaves in smoking mixtures, which have been described as very potent (Kuhnlein and Turner 1991).

The yew's hard wood is ideal for carving and takes on a polished look. Many useful implements can be made, including archery bows, clubs, paddles, digging sticks, eating utensils, snowshoe frames, and awls.

Warning Since yews are known to be toxic, the use of any part as food is not recommended.

Section III - ANGIOSPERMS: DICOTS

The dicots can be distinguished by several key characteristics. In dicots, the embryos have two cotyledons (seed leaves), and the leaves are usually net-veined. The flower parts are in 4's or 5's, rarely in 3's, and the plants are herbaceous or woody. Most importantly, the possession of cambium cells distinguishes dicotyledons from the monocotyledons.

Maple Family - ACERACEAE

This family consists of two genera and about 200 species distributed worldwide. In general, they are shrubs or trees with opposite leaves that may be simple or compound. The flowers are small, usually appearing before the leaves. The family is of economic importance as a source of timber, ornamentals, and sugar. Maple wood is considered to be heavy, tough, compact, and very hard. Its light brown color with a dense even grain and fine texture makes it one of the best woods for furniture, veneers, and flooring. It is also used in making violins, tool handles, and pianos.

Maple (*Acer*)

General Description Three native maple species are found in Idaho: Rocky Mountain maple (*A. glabrum*), bigtooth maple (*A. grandidentatum*), and box elder (*A. negundo*). Maples are deciduous trees or shrubs with male and female flowers on the same or separate plants. Flowers are arranged in racemes, corymbs, or umbels. Fruits are winged schizocarps that resemble tiny "helicopters" when blown by the wind. Maples are usually found in moist places in canyons, hills, and along streams from low to high elevations.

Field Notes/Uses Sap can be harvested in much the same way as the eastern sugar maple. To obtain sap, simply bore a small hole into the tree a couple feet above the ground. The sunny side of the tree is usually the ideal spot to bore. Insert a small grooved wooden peg into the hole. This peg will be the spigot. If the tree is ready to flow, sap will immediately begin to flow after drilling. Hang or place a container under the spigot to collect sap. After

Maple Candy

To make maple candy, boil an equal amount of maple syrup and dark corn syrup until the mixture forms balls when dropped in cold water. Drop them onto a buttered pan to form thin discs about the size of a quarter. Allow to cool, and then enjoy.

collecting sap, seal the hole to protect it from infection and further sap loss while it heals.

Next, the sap must be boiled down because the majority of it is water. Only a small fraction of the original volume collected will be left. You may boil the sap so far as your personal taste dictates. As an alternative to boiling, Hart (1996) indicates that the collected syrup can be allowed to freeze overnight, which allows the water to separate from the sap. The frozen water can be easily discarded.

The inner bark of all maples can be eaten in emergencies. A tea made from the inner bark of box elder is used to induce vomiting (Foster and Duke 1990). The young shoots of mountain maple can be used as asparagus (Harrington 1967). The winged seeds of box elder can be roasted and eaten (Thompson and Thompson 1972).

Native Americans used the young saplings for basketry work. The saplings were split into quarters, as a white wrapping or sewing strand in coiled basketry work. In some places, maple thickets were intentionally manipulated by burning the old growth to promote new growth. These straight, uniform shoots were highly valued as good basketry material. Maple wood has been used to make snowshoe framing, mush paddles, and other household utensils. Knots and burls on tree trunks can be used for making bowls, dishes, and other items. Gum from the buds was mixed with animal fat and used as a hair tonic. Inner bark can be shredded and twisted into a coarse rope.

Amaranth Family - AMARANTHACEAE

The Amaranth family contains more than 60 genera and 900 species distributed worldwide. Many are weedy species of little economic importance. Some species of *Amaranthus* are cultivated for their red pigmentation. The family name originates from the Greek *amarantos*, which means unfading, possibly alluding to the "everlasting" quality of the papery perianth parts.

Pigweed, Amaranth (*Amaranthus*)

General Description Of the approximately 50 widespread species of *Amaranthus*, eight occur in Idaho. These are *A. albus* (prostrate pigweed), *A. blitoides* (mat amaranth), *A. californicus* (California amaranth), *A. hybridus* (slim amaranth), *A. powellii* (Powell's amaranth), *A. retroflexus* (redroot amaranth), and *A. rudis* (tall amaranth). They are herbaceous annuals with small greenish flowers, and alternate entire or wavy margined leaves. Pigweeds occur in many different habitats and often hybridize, making identification difficult.

Field Notes/Uses Used by Native Americans, the dried small black seeds of *Amaranthus* have been found in many archeological remains. Seeds of all species can be eaten whole as a cereal or ground into meal, and made into cakes. The seeds are best collected in summer when the plants are fully mature. To free the seeds from their husks, rub the seed clusters between your hands. You can then winnow the seeds if there is a breeze, or if air is calm, slowly pour the seeds out of your hands and blow the chaff away. The seeds contain approximately 15 grams of protein per 100 grams, more than is found in rice and corn, and equal to, if not surpassing that found in wheat (Turner and Kuhnlein 1991). When ground into flour, *Amaranthus* has a distinctive flavor that is a bit strong used alone. We find it is better when mixed with other flours for breads and pancakes.

The highly nutritious *Amaranthus* contains more fiber and calcium than any other cereal grain in addition to a wide spectrum of vitamins and minerals, including Vitamins A and C, calcium, magnesium, and iron. *Amaranthus* is rich in the amino acid lysine, a product scarce in true cereal grains, thereby providing a more balanced source of protein (Tull 1987, Ritchie 1979).

The edible young shoots and leaves have a pleasant taste if eaten as a potherb soon after collection. Since the plants can accumulate nitrates, it is wise not to consume large quantities where nitrate fertilizers are used. Kingsbury (1964) reports that livestock losses have occurred as a result of excessive consumption of *Amaranthus* plants by the animals. The leaves of *Amaranthus* contain oxalic acid which binds with calcium restricting its absorption by the body. As long as your diet contains plenty of calcium from other sources, eating Amaranthus and other vegetables that contain oxalic acid (e.g., spinach, woodsorrel), should not cause any health problems (Tull 1987).

Amaranthus also has astringent

An Essential Amino Acid

Lysine is one of numerous amino acids that is needed for growth and repair of tissue. It is considered one of of the nine "essential" amino acids because it comes from outside sources such as foods or supplements. Like all amino acids, lysine functions as a building block for proteins and has a key role in the production of various enzymes, hormones, and disease-fighting antibodies. Though many foods supply lysine, the richest sources are red meats, fish, and dairy products. Veegetables, on the other hand, are generally a poor source of lysine with the exception of legumes (beans, peas, lentils).

properties and can be used for treating diarrhea, excessive menstrual flow, hemorrhaging, and hoarseness (Foster and Duke 1990). Angier (1978) also considers Amaranthus helpful in treating mouth and throat inflammations and sores, and in quelling dysentery and diarrhea. Steeping dried leaves in boiling water (the more leaves steeped, the stronger the tea) was considered a valuable remedy.

In the Mid-west, several species of *Amaranthus* are being grown as agricultural crops for use in cereals and bread. It is photosynthetically efficient and produces a high yield of both greens and seeds. *Amaranthus* was an important food in the past and may become an important one for the future.

Sumac Family - ANACARDIACEAE

There are approximately 79 genera and 600 species in this family. Products originating from the Sumac family include resins, oils, lacquers, edible fruits, ornamentals, and tannic acid. Although many members of this family produce edible fruits, the resinous oils can produce extreme dermatitis in sensitive individuals. The family contains the infamous poison ivy (*Toxicodendron rydbergii*), poison oak (*T. diversilobum*), and poison sumac (*T. vernix*).

Sumac (Rhus)

General Description Of the approximate 60 species in this genus that occur worldwide, two occur in Idaho. Smooth sumac (*R. glabra*) is a tall shrub with pinnately compound leaves that turn a brilliant red in fall. The flowers are greenish in cone-shaped clusters, and the fruits are berry-like, red and velvety. It often occurs in large masses and is found along dry banks, and roadsides in the foothills. Skunkbush (*R. trilobata*) is a shrub that occurs on dry sunny slopes, has compound leaves with three leaflets, of which the middle leaflet is largest. Skunkbush has small yellow-green flowers that bloom before the leaves come out. The red-orange fruits are sour tasting. The genus name comes from the Greek

Sumac-ade

The drink, called "sumac-ade," can be made from the juice of the fruit. To make this drink, simply collect several clusters of the ripened berries, clean off any excess herbage, then bruise the fruits slightly, and extract the juice by soaking them in enough cool to warm water to cover the fruit fully. Since sumacs contain high levels of tannic acid, use cool water rather than hot water so little or no tannic acid will be extracted. Soak the berries for about 20-30 minutes, strain through a cloth, and drink.

rhous, which is the name of a bushy sumac.

Field Notes/Uses The berries of both smooth sumac and skunkbush can be eaten raw, or soaked in cold water to make a refreshing drink. Malic acid (the cause of tartness in apples) flavors the sour fruits of the sumacs. Tannic acid which is present in all parts of the plant can be used in tanning leather (hides). The leaves, branches, and fruits provide colorfast dyes for wool. The stem produces a yellow dye and the berries a tan or beige dye. Tannic acid works as a natural mordant (dye fixer), as such the fibers do not need to be treated with other chemicals (Tull 1987).

The slender, flexible branches of skunkbush can be used for weaving baskets as they are somewhat vine-like. The branches can also be used as chew-sticks to clean teeth and massage gums. Take a small stem several inches long, remove the outer bark and chew on the tip to soften fibers. Since some people may have an allergic reaction to the oils of sumac, it is recommended that this be done sparingly. In the Appalachian region, the leaves of smooth sumac were smoked to treat asthma.

Poison Ivy (*Toxicodendron rydbergii*)

General Description Poison ivy is a low shrub or woody vine found in waste places, hillsides and rocky ravines in the lower elevations of the state. The leaves are compound with three green, that are oval in shape, and pointed leaflets, which turn bright red or orange in the fall. The white flowers arise from the leaf axils, and the fruits are white berries.

Field Notes/Uses The foliage of poison ivy is poisonous, causing contact dermatitis. The actual skin irritant is found in the sap. The itchy or painful rash that develops from contact with the sap is greatest in spring and summer when the sap is abundant and the plant is easily bruised. Shortly after contact, the symptoms include itching, burning, and redness. Small blisters may appear after a few to several hours. Severe dermatitis, with large blisters and local swelling, can remain for several days and may require hospitalization.

Urushiol

The heavy oil found in all parts of poison ivy is called urushiol. It is incredibly potent, and an amount the size of a pinhead can cause rashes and blisters in sensitive people. It is also long-lasting, so much so that botanists handling four hundred year old specimens in collections have developed rashes. Urushiol is confined to the canals inside the plant's leaves, stems, and roots, and it oozes out readily when bruised or chewed on. It takes only 10 minutes to penetrate skin, so washing quickly with soap and water is the simplest field remedy.

Secondary infections may occur when the blisters are broken. To help alleviate itching immediately, thorough washing with soap and water after contact is recommended.

Since droplets of the irritating chemical can be carried in smoke on dust particles and ash, do not burn poison ivy. Smoke carries the oil and can produce a rash over the whole body. If inhaled, a rash can develop in the throat, bronchial tubes, and lungs. The oil can even spread through the body in the blood stream. Although poison ivy can cause havoc for humans, robins, cedar waxwings, flickers, woodpeckers, and other birds relish the berries.

Carrot Family - APIACEAE

Approximately 300 genera and 3,000 species are found in this family. About one-quarter of the genera and 10% of the species are native to the United States. The Carrot family is of considerable economic importance because of numerous food plants, condiments, ornamentals, and poisonous species. Some familiar members of this family include Carrot (*Daucus*), Parsnip  (*Pastinaca*), Celery (*Apium*), Anise (*Pimpinella*), and Parsley (*Petroselinum*). No wild members of this family should be eaten until they have been accurately identified. Correct identification usually requires the mature fruit called a schizocarp (a dry fruit that splits into two halves).

Angelica (*Angelica*)

General Description The five species that occur in Idaho can be found along streams and moist soil in the mountains. In general, they are tall, often aromatic, and taprooted perennials with leaves that are mostly twice divided into broad, toothed or lobed leaflets. The flowers are mostly white. The fruit is narrowly elliptical to nearly orbicular with winged ribs on the outer face.

Field Notes/Uses While there are a number of edible species of angelica, we have not found any information regarding the edibility of the five species that occur in Idaho: *A. arguta* (Lyall's angelica), *A. dawsonii* (Dawson's angelica), *A. kingii* (King's angelica), *A. pinnata* (small-leaf angelica), and *A. roseana* (rose angelica). Therefore, the internal use of angelica species in Idaho is not recommended until studies have been conducted concerning their toxicity, and because they superficially resemble the poisonous water hemlock (*Cicuta maculata*). All species of angelica contain furanocoumarins, which increase skin photosensitivity and may cause dermatitis.

Very little is known about the medicinal aspects of Idaho species. Tilford (1993) describes Angelica, in general, as an antispasmodic, expectorant, diaphoretic, diuretic, an effective astringent to the stomach lining, and a menstrual stimulant that helps reduce cramps. Angier (1978) describes a volatile oil in eastern species of angelica that was used to treat colic and digestive gas. Poultices from the mashed roots of angelica were applied for arthritis, chest discomfort, and pneumonia.

Warning Angelica closely resembles the highly poisonous water hemlock (*Cicuta douglasii*). Positive identification of the plants is paramount. The identification of young plants of angelica and water hemlock can be made by examining the leaf venation. The leaf edges of angelica are serrate and pinnately divided into opposing pairs, like water hemlock, but the leaf veins extend from the midribs to the outer tips of the serrations. Water hemlock has leaf veins terminating within the notches of the serrations.

Wild Celery (*Apium graveolens*)

General Description This is a stout smooth perennial with large pinnate leaves that have few leaflets. The leaves are deeply 3-lobed and circular in shape. The celery stalk is familiar to all. The plants are found in moist soil at the lower elevations.

Field Notes/Uses This is the common celery which has escaped cultivation in some areas. Be warned, this species can absorb dangerous levels of nitrates in its tops. It is best to discard the tops.

Cutleaf Waterparsnip (*Berula erecta*)

General Description This is a glabrous, aquatic or semi-aquatic perennial with fibrous roots and leafy branched stems. The submerged leaves are often dissected into many hair-like segments; emersed leaves are pinnately divided. Flowers are borne in compound umbels with 6-15 rays and green linear bracts at the base of both the primary and secondary umbels. Petals are white and the nearly round fruit has low ribs on the face. Waterparsnip occurs in shallow water of ditches, sloughs, and streams.

Field Notes/Uses Apache in the southwest supposedly used the leaves and flowers for food.

American Thorow Wax (*Bupleurum americanum*)

General Description This is a yellow-flowered plant with undivided leaves (a rather unusual trait for a member of the Carrot family), but are linear and toothless. There is a leaf-like bract at the base of each small, flower head and again at the base of the main cluster.

Field Notes/Uses This is a genus of 70-100 species of annual and perennial herbs and shrubs with simple leaves and umbels of typically yellow flowers. Several species are grown as ornamentals. In China, thorow wax plants (often in combination with other plants) were used in relieving the side effects of steroids.

Caraway (*Carum carvi*)

General Description This glabrous biennial has a taproot and slender, leafy stems and grows 1 to 3 feet tall. The basal leaves have petioles and blades 3 to 7 inches long, and are pinnately divided into highly dissected and feather-like segments. The stem leaves are smaller with wider sheaths. The white flowers occur in numerous compound umbels comprised of 7-14 rays. The fruits are elliptical to almost round in shape and has low ribs on the face.

Field Notes/Uses Caraway seed has been a popular and widely-used spice for several thousand years. The plant is native to the Mediterranean regions of Europe, Asia and North Africa, and is additionally cultivated in Northern Europe, Russia and the United States though for many years the Netherlands has been the main supplier in world trade, with up to thousands of acres under cultivation. The seeds are used for spicing a wide variety of bakery products, cooked meats, salads, pickles and drinks: the characteristic flavor of caraway derives from the essential oil, which is extracted from small channels within the seeds. This oil is valued in pharmacology as an antispasmodic and carminative; it is used in perfumery and also in preparations such as gargles and mouthwashes, having some antibacterial properties. More recently it has been discovered that the largest constituent of the oil, carvone, has potential uses as an insect repellent, as a suppressant of sprouting in stored potatoes, and for inhibiting the growth of some fungi.

Water Hemlock (*Cicuta*)

General Description Three species of *Cicuta* can be found in Idaho: *C. bulbifera* (bulbet-bearing water hemlock), *C. douglasii* (western water hemlock), and *C. maculata* (spotted water hemlock). They can be found in marshes and along the edges of streams and ponds from low to mid-elevations. Water hemlock is a stout perennial from fleshy, fascicled roots. Leaves are 1-2 times pinnately divided into narrowly lance-shaped, sharply toothed leaflets. The veins of the leaflets terminate at the notches between the teeth (see photo). Numerous white to greenish flowers are arranged in compound umbels. Fruits are slightly flattened with thickened ribs on the faces. The bruised foliage produces a musky odor.

Field Notes/Uses These is an extremely poisonous plant if ingested. In fact, water hemlock has been described as the most violently poisonous vascular plant in North America. The whole plant contains cicutoxin, a resin-like substance that depresses the respiratory system, with the root being particularly dangerous. A single mouthful of the plant is capable of killing an adult. Water hemlock has been used throughout the ages to execute criminals and kings. Many children have been fatally poisoned by blowing into whistles made from hollow stems of Water Hemlock. In Oregon, Native Americans soaked arrows in Cicuta juice, rattlesnake venom, and decayed deer liver to poison arrow tips for hunting (Davis 1952).

Strike (1994) indicates that water hemlock roots were mashed and smeared on a hot stone to relieve pain in sore arms or legs. The mashed root was then pressed against the sore arm or leg.

Following is a graphic description from Jacobson (1915) of poisoning due to the ingestion of water hemlock in Europe. If anything, it should instill into the minds of most wild food gatherers the need to positively identify a plant species before eating it.

> "When about the end of March 1670, the cattle were being led from the village to water at the spring, in treading the river banks they exposed the roots of this *Cicuta* (water hemlock) whose stems and leaf buds were now coming forth. At that time two boys and six girls, a little before noon, ran out to the spring and the meadow through which the river flows, and seeing a root and thinking that was a golden parsnip, not through the bidding of any evil appetite, but at the behest of wayward frolicsomeness, ate greedily of it, and certain of the girls among them commended the root to others for its sweetness and pleasantness, wherefore the boys, especially, ate quite abundantly of it and joyfully hastened home; and one of the girls tearfully complained to her mother she had been supplied meagerly by her comrades, with the root.
>
> Jacob Maeder, a boy of six years, possessed of white locks, and delicate though active, returned home happy and smiling, as if things had gone well. A little while afterwards he complained of pain in his abdomen, and, scarcely uttering a word, fell prostrate to the ground, and urinated with great violence to the height of a man. Presently he was a terrible sight to see, being

seized with convulsions, with the loss of all his senses. His mouth was shut most tightly so that it could not be opened by any means. He grated his teeth; he twisted his eyes about strangely and blood flowed from his ears. In the region of his abdomen a certain swollen body of the size of a man's fist struck the hand of the afflicted father with the greatest force, particularly in the neighborhood of the ensiform cartilage. He frequently hiccupped; at times he seemed to be about to vomit, but he could force nothing from his mouth, which was most tightly closed. He tossed his limbs about marvelously and twisted them; frequently his head was drawn backward and his whole back was curved in the form of a bow, so that a small child could have crept beneath him in the space between his back and the bed without touching him. When the convulsions ceased momentarily, he implored the assistance of his mother. Presently, when they returned with equal violence, he could not be aroused by no pinching, by no talking, or by no other means, until his strength failed and he grew pale; and when a hand was placed on his breast he breathed his last. These symptoms continued scarcely beyond a half hour. After his death, his abdomen and face swelled without lividness except that a little was noticeable about the eyes. From the mouth of the corpse even to the hour of his burial green froth flowed very abundantly, and although it was wiped away frequently by his grieving father, nevertheless new froth soon took its place."

Poison Hemlock (*Conium maculatum*)

General Description Poison hemlock is a biennial with a stout taproot and a disagreeable odor when crushed. The stems are purple-blotched and hollow, and the leaves are pinnately dissected with a lacy appearance to them. The flowers are white in compound umbels, and the fruits are egg-shaped, flattened with prominent, wavy ribs. The plant is usually found in disturbed sites and waste places at low elevations.

Field Notes/Uses This is an extremely poisonous plant. Death is said to result from the ingestion of the leaves, roots or seeds. The most famous use of poison hemlock was by the ancient Greeks as a humane method of capital punishment. It is said to be quite painless and the recipient's mind remains clear to the end. Introduced from Europe, poison hemlock has established itself as a common weed. Socrates was said to be killed by the plant in 399 BC when he was forced to drink it.

Spring Parsley (*Cymopterus*)

General Description The nine species of *Cymopterus* in Idaho are low perennial herbs with long, thick taproots. The leaves are 2-4 times pinnately divided into small ultimate segments, and the flowers are yellow or white in terminal, compound umbels. The round fruits have winged ribs on the outer faces. Most of the species occur in dry soils or gravelly slopes at the lower elevations.

Field Notes/Uses All species produce edible roots. We found the older roots more fibrous than the younger ones. The root can be used in stews or it can be boiled or roasted in a pit, mashed and dried into cakes. When dried, it will keep indefinitely. During the Lewis and Clark expedition this was known as *kouse* (bread of cows). The old roots can also be used as an effective insect repellant when boiled. Just sprinkle the tea around camp and in sleeping areas (Sweet 1976, Strike 1994).

The upper parts of the plants have been used raw or as potherbs. If cooked, they will require several changes of water. The seeds of some species are edible when ground and used as flour (Olsen 1990).

Queen Anne's Lace, Wild Carrot (*Daucus*)

General Description Two species occur in Idaho. These plants resemble poison hemlock (*Conium maculatum*), but are readily distinguished from each other in that *Daucus* has stems and leaves that are distinctly hairy.

Field Notes/Uses *Daucus carota* (Queen Anne's Lace) is common at the lower elevations on roadsides, fields, pastures, waste places, and moist clearings. Introduced from Europe, it is the ancestor of the cultivated carrot. The first year's roots can be prepared like garden carrots. We found the older roots tough and stringy. The roots can also be dried and roasted and then ground for use as a coffee substitute. The plant was used extensively by many Native Americans and should be kept in mind as an emergency food.

A tea made from the root has been traditionally used as a diuretic to prevent and eliminate urinary stones and worms. Laboratory studies confirm the bactericidal, diuretic, and worm expelling properties (Foster and Duke 1990). The seeds of this species are also considered to be a folk "morning after" contraceptive.

A native annual, *Daucus pusillus* (American wild carrot) is more slender than *D. carota*, and has fewer flowers per umbellet. It is a species of dry, open, rocky areas or grassy sites at lower elevations. Crushed seeds have been used as a contraceptive and herbal "morning after pill" for at least 2,500 years (Pojar and MacKinnon 1994). American wild carrot has been shown to be successful for use as a contraceptive in laboratory trials, and is used in some areas today.

Costanoan Indians in California used *D. pusillus* to reduce fevers, heal snake bites, cure colds, and as a general blood medicine. As a poultice, the plant was used on bruises and swellings (Strike 1994).

Warning Queen Anne's lace and wild carrot resemble poison hemlock (*Conium maculatum*). They may cause dermatitis and blisters. They are also not recommended for use as contraceptives.

Cowparsnip (*Heracleum maximum*)

General Description Cowparsnip is a stout perennial up to or more than seven feet tall. The lower leaves are three lobed, resembling a maple leaf up to 14 inches long. The white flowers are in compound umbels and the fruits are egg-shaped with only the marginal ribs winged. It is usually found in moist soils around streams, seeps, and avalanche chutes up to subalpine environments. The genus is named for Hercules, who is reputed to have used it as a medicine.

Field Notes/Uses The young stems of cowparsnip can be peeled and eaten raw, but are best when cooked. The hollow base of the plant can be cut into short lengths and used as a substitute for salt by eating or cooking with other foods. The young leaves are also edible after cooking, but we find them not very tasty. The leaves can also be dried and burned, and the ashes used as a salt substitute. Strong and bitter tasting, the cooked roots are said to be good for digestion, as well as for relieving gas, and cramps (Willard 1992). In our experience, some plants are much more palatable than others.

The seeds can be sparingly added to salads for seasoning. However, you should be aware that the mature, green seeds have a mild anesthetic action on tissues in the mouth. When gently chewed and sucked, they will numb the tongue and gums in a manner similar to clove oil.

Medicinally, root pieces were placed in tooth cavities to stop pain (Tilford 1993). An infusion for a sore throat can be made by soaking the mashed root in water (Willard 1992). Root tea was used for colic, cramps, headaches, sore throats, and flu (Moore 1979).

The leaves of cowparsnip are large enough to be used as a toilet paper substitute and as a mild insect repellent. However, since furanocoumarin is present in the sap and the outer hairs, it may be a problem for people with sensitive skin. When the sap comes in contact with the skin in sensitive people it causes a type of "sunburn" effect (i.e., redness, blistering, and running sores) when exposed to light. As a poultice, the leaves were used externally for sores, bruises, and swellings (Foster and Duke 1990).

The older stems, before the flower cluster unfolds, can be peeled and the inner tissue eaten raw or cooked. While it is edible, it does have an unpleasant taste. Cooking it in a couple changes of water usually helps the taste and digestibility. In any case, cowparsnip is considered to be an excellent survival plant in the mountains. Strike (1994) indicates that cowparsnip has been probably the most intensively used springtime green among Native Americans in Canada.

Warning Do not confuse this plant with other species in the same family that are highly toxic (i.e., *Cicuta* and *Conium maculatum*).

Licorice Root (*Ligusticum*)

General Description The five species of *Ligusticum* in Idaho are characterized as perennials with taproots sheathed by old leaf bases at the crown. The leaves are 1-3 times pinnately dissected, the white to pinkish flowers are arranged in compound umbels. Fruits are oblong to elliptical with winged ribs. They are generally found in moist habitats in the foothills and mountains

Field Notes/Uses Some of the species contain alkaloids, but the green stems and roots of *L. filicinum* (fernleaf licoriceroot) may be eaten raw or cooked. Additionally, the tender leaves of *L. grayi* (Gray's licorice-root) were soaked in water, cooked, and used as a meat substitute when acorns were eaten. Several different species have been used medicinally. They contain volatile and fixed oils, and a very bitter alkaloid that has been shown to increase blood flow to coronary arteries and the brain.

Tilford (1993) describes *L. canbyi* (Canby's licoriceroot) as an antiviral, expectorant, diaphoretic, anesthetic to the throat, and useful in the treatment of upper respiratory infections. Native Americans commonly chewed the dried roots of *L. canbyi* for relief from sore throats, colds, toothache, headache, stomachache, fever, and heart problems (Hart 1996).

Desert Parsley, Biscuitroot (*Lomatium*)

General Description Many species of biscuitroot occur in Idaho. They are perennial plants with thick roots and leaves that are divided several times from the base. The white, yellow, pink, or purplish flowers are in compound umbels. The fruits are flattened and elliptical to oval in shape, and the margins may or may not be winged. Most species are found in dry ground or rocky situations. The genus name means "small border," alluding to the wings of the fruit.

Field Notes/Uses All biscuitroot species have edible roots and were an important staple among many Native Americans. They can be eaten raw or cooked, or dried and ground into flour. The flour can then be kneaded into dough, flattened into cakes, and dried in the sun or baked. Some of the species we have tried were too resinous to enjoy. Personal taste will guide one to choose the more palatable species.

The green stems can be eaten after boiling in the springtime, but as summer progresses they become tough and fibrous. A tea can be brewed from leaves, stems, and flowers. The tiny seeds nutritious raw or roasted, can be dried and ground into meal.

The plants are rich in Vitamin C. Seeds were chewed for colds and sore throats, and sap from the roots was used to treat cuts and sores (Moore 1979). A poultice of pulverized roots was applied to a newborn baby's umbilical cord to facilitate healing. Roots were also chewed to relieve sore throats (Strike 1994). Recent studies have shown that *L. dissectum* (fernleaf biscuitroot) has an ability to kill certain forms of influenza virus, especially those that infect the respiratory tract. It also has other anti-microbial and immuno-stimulating qualities (Tilford 1993).

Caution As with any member of the Carrot family, positive identification is important before consumption. Strike (1994) indicates that some indigenous peoples considered the mature stalks, leaves, roots, and flowers of L. dissectum as poisonous. In fact, the roots were used as a fish poison and insecticide by some Native People in the Columbia Plateau. The plant contains phototoxic compounds of the furanocoumarin group and one or more of these compounds is responsible for the fish poison and insecticidal properties found in the chocolate tipped roots.

Leafy Wild Parsley (*Musineon divaricatum*)

General Description This is perennial from a purple taproot. Plants are about four to eight inches tall. Stems spread out from the rootcrown and then curve upward. The stalked leaves are thick and glossy, and divided two or three times into leaflets and lobes. The tiny yellow flowers are borne in small stalked clusters atop a main stalk. A whole flat-topped group of flower clusters is about one or two inches wide. The fruits are egg-shaped and evidently ribbed, but without wings.

Field Notes/Uses The genus is similar to *Lomatium* and *Cymopterus*, but is distinguished by the completely wingless fruit. The generic name *Musineon* is surely a corruption, possibly of the Greek *mouseion* 'museum' or 'shrine of Muses.' Leafy wild parsley was first described for science by the eminent German botanist Frederick Pursh (1774-1820). While there are no known economic uses for this species, the fleshy root of leafy wild-parsley was used for food, raw or cooked, by the Blackfoot and Crow Indians.

Indian Potato, Snow Peas (*Orogenia*)

General Description Two species, *O. fusiformis* (California Indian potato) and *O. linearifolia* (Great Basin Indian potato), occur in Idaho. They are small perennial plants with fleshy roots. Flowers are white and in compound umbels. Look for the species soon after snows melt in the mountains in spring and early summer. They are sometimes called Snow Drops.

Field Notes/Uses The roots of both species can be boiled, steamed, roasted or baked in any way used for preparing potatoes. When small, they can be eaten raw. They can also be cooked and mashed into cakes for drying, and when protected from moisture, will keep a long time. The hard cakes can be soaked and cooked in stews.

Sweetroot (*Osmorhiza*)

General Description The four species of sweetroot in Idaho are herbaceous perennials from stout roots, with leaves twice divided into 3's. The flowers are borne in open, compound umbels that arise from leaf axils. The fruit is spindle-shaped and compressed along the sides. In general, they occur on moist slopes, open areas, and forests.

Field Notes/Uses The roots of *O. occidentalis* (western sweetroot) taste and smell like licorice or anise, and can overwhelm the taste buds if eaten in large amounts. In small quantities, Western sweetroot can liven up the taste of teas (or meals) that are otherwise bland or unpleasant.

The leaves of *O. berteroi* (sweetcicely), known as "dryland" parsnip, can be boiled and then eaten. The roots were dug in the spring and either pit cooked or boiled as a vegetable. To us the taste is reminiscent of baby carrots.

Uses of related species include a poultice from roots for boils, cuts, sores, and wounds, and root tea for sore throat and upset stomach (Foster and Duke 1990).

Caution Western sweetroot resembles the very poisonous water hemlock (*Cicuta douglasii*), but the strong smell of anise gives it away as sweetroot. Also, the venation of water hemlock is unique among the Apiaceae (Carrot family). Sweetcicely can be confused with baneberry (*Actea rubra*). However, baneberry is easy to distinguish from sweetcicely by the cluster of red or white berries.

Wild Parsnip (*Pastinaca sativa*)

General Description Wild parsnip is a biennial with stout, leafy stems that arises from a large taproot. The basal leaves are once pinnately compound into usually 9-13 lance- to egg-shaped leaflets. Flowers are yellow and in compound umbels. The fruits have fine ribs on the outer face. This introduced plant from Europe is the wild form of the cultivated parsnip and can be prepared and eaten in the same way. It can be found in damp disturbed areas at the lower elevations.

Field Notes/Uses The taproot from first year plants can be eaten raw or boiled until tender. The root of the plant was used by many people, from the ancient Romans to Native Americans. In Holland, the plant was used in soups. The Irish made a beer by boiling the roots with water and hops, and then allowed the mixture to ferment. A tea from the roots was used by Native Americans to treat sharp pains. The roots were also poulticed and used on inflammations and sores.

Caution Due to the presence of xanthotoxin, the plant may cause photodermatitis. The symptoms are much like those from exposure to poison ivy, but of longer duration. Xanthotoxin is used to treat psoriasis and virtiligo. Therefore, avoid contact and exposure to sunlight.

Yampah (*Perideridia*)

General Description *Perideridia bolanderi* (Bolander's yampah) and *P. gairdneri* (common yampah) are found in moist soils of meadows and other open

 areas in Idaho. They are perennials from tuberous roots with compound leaves. The flowers are white and the fruits are usually flattened laterally.

Field Notes/Uses All of the species within this genus are essentially edible. They were an important food of many indigenous peoples from British Columbia to California and Great Basin region. The raw finger-like roots have a pleasant, nutty flavor when eaten raw, and resemble carrots when cooked. They are best when dug up before the flowers appear. The roots should be washed and peeled before cooking.

They can be easily dried and will keep well for future use. When dried, the roots can be pounded and ground into flour or mashed into cakes. The seeds may be parched and ground or eaten whole.

Snakeroot (*Sanicula*)

General Description Two species, *S. graveolens* (northern sanicle) and *S. marilandica* (Maryland sanicle) can be found growing at the lower elevations in Idaho. Both are erect perennials with 3-5 lobed leaves. The flowers are in compound umbels and the fruits are flattened laterally, densely covered by bristles.

Field Notes/Uses The herbage of both species contain various alkaloids and should therefore be regarded as inedible. The root of Maryland sanicle was used in teas for menstrual irregularities, pain, and rheumatism fevers, and as a poultice for snakebites (Foster and Duke 1990).

Hemlock Waterparsnip (*Sium suave*)

General Description This is the only species of the genus in Idaho. It is a stout plant up to five feet tall. The leaves are pinnately divided, the flowers are white, and the fruit is oval in shape. It is usually found in water or swampy areas in the mountains.

Field Notes/Uses The long fleshy root of hemlock waterparsnip, which is edible raw or cooked, has a sweet, carrot-like taste. The leaves and younger stems are also edible after cooking, but we found them better when boiled until tender. The older plants and flowers should be avoided because they are toxic and have been suspected of poisoning a wide range of livestock.

Warning The plant is very similar in form and habitat to *Cicuta maculata* (spotted water hemlock) which is the most poisonous vascular plant in North America (Kingsbury 1964). Both plants produce white flowers in umbrella-like clusters and both grow in swampy ground at lake or pond edges. Water Parsnip has leaves that are once-compound, whereas the leaves of water hemlock are 3-times compound. Water hemlock also has a distinctive turnip-like swelling at the base of the stem, which is usually chambered when cut open vertically and exudes a yellowish liquid along the cut surface. Therefore, when in doubt, leave it alone!!!

Woollyhead Parsnip, Ranger's Button (*Sphenosciadium capitellatum*)

General Description This is a stout plant with large leaves. The leaflets are linear-oblong to ovate-lanceolate, usually with serrate to dentate margins.

Woollyhead parsnip can be found in moist or wet meadows, bogs or streambanks to the middle elevations.

Field Notes/Uses The roots were chewed by Native Americans in California to relieve sore throats, and a root decoction was used to treat bronchial problems (Strike 1994). An infusion of the roots was used by the Maidu and Paiute Indians to repel lice (Strike 1994).

Meadow Zizia (*Zizia aptera*)

General Description The plant has several clusters of small bright yellow flowers growing in an open mounded form with lush, heart shaped foliage. This plant has several common names including Golden Alexander, Heartleaf Alexander, Meadow Zizia and Heart-leaved Meadow Parsnip.

Field Notes/Uses Non-reproductive plants form compact rosettes up to 6 inches in diameter. Reproductive plants may attain heights of 3 feet, and produce compound umbels of small, bright yellow flowers from May to July. These protogynous flowers are pollinated by a variety of bees and flies, some species of which focus exclusively on *Zizia* species. Although the plant produces defensive compounds, including the unique furanocoumarin "apterin" (named for the species), a number of insect species are herbivores on leaves, stems, and seeds of *Zizia aptera*.

The presence of secondary compounds such as apterin may contribute to the potential medicinal value of *Zizia* species. For example, *Z. aurea* roots have been used by Native Americans as a tea to cure fevers, and the plant has been referred to as a vulnerary (wound-healing) agent (Foster and Duke 1990). This species has also been used to induce sleep and for alleviating syphilis (Foster and Duke 1990). However, the specific medicinal properties of *Z. aptera* have not been documented.

Dogbane Family - APOCYNACEAE

This is a large family of about 200 genera and 2,000 species that are mostly found in the tropics. Nearly all of the members within this family are poisonous and usually have milky juice. Some of the well-known genera are ornamentals such as *Vinca minor* (periwinkle) and *Nerium oleander* (oleander). In recent years *Rauwolfia serpentina* (Indian snakeroot), a tropical tree, was found to yield a wonder drug used in the treatment of high blood pressure.

Dogbane, Indian Hemp (*Apocynum*)

General Description The two species in Idaho include spreading dogbane (*A. androsaemifoilium*) and Indian hemp (*A. cannabinum*). They are perennial herbs with milky juice, have leaves that are opposite, and the pink, bell-shaped flowers are borne in cymes. There is considerable hybridization between species. Apocynum is Greek meaning "noxious to dogs."

Field Notes/Uses The primary use of dogbanes is for fiber. The stem fibers are strong and can be used for rope making, mats, baskets, bowstrings, fishing lines and nets, sewing, animal trap triggers, snares, cordage for bow and drill fire making, and general weaving. One of the easiest ways to isolate the fibers is to soak the stems in water. Archeologists in Utah have discovered nets made with the fiber dating back to about 5000 B.C. Many Native American tribes used dogbane to make rabbit-catching nets.  Some of these nets were about 200 feet long, three to four feet high, with a three inch opening. The nets were propped on sticks across level ground. The men formed a line some distance away and advanced toward the nets, beating the brush with sticks, and driving the rabbits into the net (Strike 1994).

Dogbanes should be considered poisonous to humans if ingested. However, Weedon (1996), Harrington (1967), and Sweet (1976) indicate that the small seeds can be parched, ground into a meal to make fried cakes. Strike (1994) believes the seeds eaten were that of *A. pumilum* (mountain dogbane); *A. pumilum* is a subspecies of *A. androsaemifolium*.

Dogbanes were extensively used as medicine by aboriginal peoples (Willard 1992, Erichsen-Brown 1989, Gunther 1988, Moore 1979). They contain highly toxic glycosides and resins with cymarin and apocannocide being major medicinal constituents found throughout the plants. Research on the compound indicates that the cardiac glycosides may be useful in treating malignant tumors. The Blackfoot Indians apparently boiled the roots of *A. cannabinum* in water for a tea for use as laxative. Millspaugh (1974) describes *A. androsaemifolium* as an emetic without causing nausea, cathartic, and quite powerful diuretic and sudorific; it is also an expectorant and antisyphilitic.

Ginseng Family - ARALIACEAE

There are about 50 genera and 500 species worldwide. Though they are of limited economic importance, several species are used as ornamentals. One species, American ginseng (Panax quinquefolius) is the famous medicinal panacea and "mind enhancer."

Wild Sarsaparilla (*Aralia nudicaulis*)

General Description This is a perennial forb found in moist woods and thickets. The leaves are solitary, up to 20 inches tall, and are twice divided, first into 3's and then into 3-5 elliptic, toothed leaflets. The flowering stem is much shorter than the leaves. The inflorescence is made up of 3-7 slender, stalked spherical clusters of numerous yellow-green flowers. The berries are purplish black. Wild sarsaparilla is indicative of relatively warm, moist woods, and is usually confined to the lower montane elevations.

Field Notes/Uses A poultice from the rhizomes can be used to treat burns and sores (Willard 1992). As a tonic, it was a diuretic that lowers fever. As a tea, wild sarsaparilla is rather pleasant tasting. Internally, it was used for coughs and purifying blood. The long rootstalk is often used as an ingredient of rootbeer. Fernald and Kinsey (1958) indicate that Native Americans relied for long periods of time on these roots during wars or when hunting. It has been used as a substitute for true sarsaparilla (*Smilax officinalis*). The roots and rhizomes of a related species, *A. racemosa*, and not found in Idaho, have been used to treat rheumatism, coughs, and backaches.

Devil's Club (*Oplopanax horridus*)

General Description The entire plant is covered with brittle, sharp yellowish spines that tend to break off in the flesh of people attempting to handle it. The white flowers are small, and are borne in a terminal, umbrella-shaped inflorescence. The fruits are red berries. Devil's club is found in moist, dense woods, usually in old-growth stands.

Field Notes/Uses When the plants are very young (when their spines are soft and harmless), they are edible. A word of warning, though, these plants tend to arm themselves very early in life.

Many herbalists believe that devil's club is a substitute for the American ginseng (*Panax quinquefolius*), however, this is not true. While it does share some of the pharmacological and therapeutic similarities with ginseng, it is not the same medicine. Devil's club was used by Native Americans for diabetes and to treat cancer.

Birthwort Family - ARISTOLOCHIACEAE

Only one genus and species occurs in Idaho, but about six genera and 400 species can be found in the tropics. Plants in this family are herbs or woody vines with commonly heart-shaped, entire leaves. The flowers often lack petals, and may be carrion scented.

British Columbia Wildginger (*Asarum caudatum*)

General Description Wildginger is a perennial plant with shiny heart-shaped leaves that forms dense mats on shaded forest floor and meadows. It is fairly common at the lower to middle elevations. The flowers are brownish purple and bell-shaped, and are borne in the axil of the leaf. The petal-like sepals are broad at the base and taper into slender threads. *Caudatum* means tail-like.

Field Notes/Uses When crushed, the entire plant has a strong lemon-ginger smell. The roots can be eaten raw or dried and ground as a ginger substitute or used to make a tea. The rootstock may also be dried and kept for later use, or candied by tenderizing the short pieces in boiling water and then boiling the pieces in heavy syrup.

A tea made from the root was drunk for stomach pains and to relieve gas (Tilford 1993). A poultice was made for headaches, intestinal pains, and knee pains. A fine powder of the dried roots was inhaled like snuff to relieve aching head and eyes (Sweet 1975). Wildginger is said to have antibiotic properties. Some Native Americans boiled the leaves, crushed and put them in bath water or rubbed them directly on the painful limb (Pojar and MacKinnon 1994, Strike 1994).

Warning The species contains asarone, a compound found in laboratory tests to cause tumors in rats, but it may not have the same effect on humans (Miller 1973). Additionally, wildginger is usually found in habitats that are slowly being destroyed by human activities. Over harvesting the plant is becoming a real concern.

Milkweed Family - ASCLEPIADACEAE

About 250 genera and 2,000 species are found in this family worldwide. Milky sap in the stems, leaves, and flowers inspired the common name for the

Milkweed family. The family is of moderate economic importance as a source of ornamentals, latex, fibers, poisonous plants, and a few food plants. The sap contains latex, and in a few species, it may yield industrially important hydrocarbons. The flowers are 5-parted. Pollination involves an insect literally pulling up the pollen mass from the anthers and directly depositing it on the stigma.

Milkweed (*Asclepias*)

General Description At least six species of *Asclepias* are known to occur in Idaho. In general, they are herbs found at the lower elevations, often in fields, along roadsides, along streams, and on hillsides. The name *Asclepias* refers to *Asklepios*, the Greek god of medicine.

Field Notes/Uses Many books on edible plants in the United States list milkweeds as an edible plant. It should be noted that in most cases they are

Wilderness Cordage

The survival and continued existence of primitive humans was as much dependent on fiber as on food. The cordage made from the fibers of wild plants can used to make blankets, sandals, baskets, clothing, nets for fishing, and snares for capturing small game animals. In a wilderness situation, you will be surprised how important a piece of string or cordage can be. There are many species of plants in Idaho that have fiber in the stem, leaves, or bark that can in one way or another be used as cordage. Some of the species discussed in this handbook include milkweed, dogbane, sagebrush, cottonwood, willow, juniper, thistle, sunflower, and nettle. There are probably other species that can be used, and finding out which ones will be a matter of experimentation.

Twisting cordage is relatively easy once the fiber has been extracted from the plant. In most cases, the fiber is located in the outer part of the plant stem. The fibers can be removed by rubbing the stem between the hands or by carefully pounding it with a rounded rock or mallet. It is important to not to break the length of the fiber while doing this. This should result in soft, thread-like fibers.

To twist the fiber into a short piece of cordage, simply roll the length of fiber down your leg with an open palm until it is rounded and reasonably uniform in diameter. However, if longer cordage is required to be used as fishing line, sewing, nets, or clothing, it will be necessary for you to twist and splice. The following directions are for right-handed people. If you are left handed, simply reverse the process.

With the strand of fiber in your left hand, bend it in half. There should be two uneven lengths hanging down. Pinch the loop you've created with the left thumb and forefinger. With your right hand, grab the strands on the outside, twist it in the outward direction about half-way, and then lay it over the inside strand. Move your left thumb and forefinger down to hold it together. Again, with your right hand, grab the new outside strand, twist it out, and again lay it over the inside strand and reposition your left thumb and forefinger down to hold it together. Repeat this a few more times. When you are about two to three inches from the end of the shorter strand, take another length of fiber and lay it on the shorter piece and twist it as though it were part of the shorter strand. Continue twisting as before until you reach the 2 or 3 inch mark with the other strand. Again, attach a new strand and twist it as part of the new one. This is called splicing. If you are doing this for the first time, you'll soon realize that you have muscles in your fingers you never knew you had.

referring to the eastern species of *A. syriaca* which does not occur in Idaho. A couple of species in Idaho that have been used as food include showy milkweed (*A. speciosa*) and spider milkweed (*A. asperula*). These latter two species also contain the cardiac glycosides that can cause severe poisoning if not properly prepared or cooked.

Harrington (1967) suggests gathering plants when they are six inches tall and then boiling for 15-20 minutes in at least 2-3 changes of water. We have tried 5 to 7 changes of water and the plants were still bitter!! The unopened flower buds can be served like broccoli by boiling in at least three changes of water.

A strong fiber can be obtained from the inner bark to make rope, fishing line, clothing, and nets. Archeologists have discovered clothing that was made from the fibers more than 10,000 years ago (Tull 1987). The silky floss found in mature Milkweed seed pods were used in making candlewicks, and the fiber can be spun like cotton. The floss is buoyant and water resistant and makes a good insulator. During World War II, schoolchildren in Canada harvested milkweed floss from the wild for the United States Navy's use as a substitute for kapok in life vests (Tull 1987). The dried pods were used as utensils and the sap was used as an adhesive.

Milkweeds contain asclepain in their plant parts and sap. Asclepain is a proteolytic enzyme that gives credence to the old pioneer remedy of applying the white juice daily to get rid of warts (Kirkpatrick 1992). However, some Native American tribes used to collect the milk of *A. speciosa* and roll it in hand until it became firm enough to chew as gum, but it was not swallowed.

Milkweeds have been used in folk medicine for hundreds of years. The powdered root of several species was reported to have been used to treat wounds, pulmonary diseases, rheumatism, and gastrointestinal problems, among other ailments. Many modern medicines were originally derived from poisonous plants. Perhaps research will validate some of the medicinal uses of milkweeds and provide us with new medicines from the old (Lewis and Elvin-Lewis 1977).

Additionally, milkweeds have the potential to furnish an exciting array of products for industry and home. In the future, as petroleum products dwindle, perhaps we will find ourselves taking a closer look at the possibilities of cultivating milkweeds for fiber, hydrocarbons, and medicines.

Warning Milkweeds can be confused with other plants producing milky juice such as dogbane (*Apocynum*), which is also considered to be poisonous. Additionally, some species of milkweed at certain stages are poisonous to animals and could affect humans when eaten raw.

Sunflower Family - ASTERACEAE

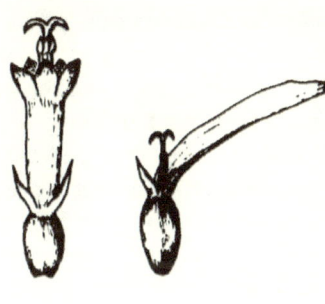

This is a very diverse family with over 20,000 species. The Sunflower family (often referred to as composites) is the second largest plant family in the world. The Sunflower family contains many economically important crop plants such as sunflowers, lettuce, and artichokes. Numerous edible and useful composites are found in Idaho.

While the family is considered by many botanists as a "difficult" group, composites, in general are relatively easy to recognize. The small flowers are arranged in heads that at first appear to be an individual flower, although it may actually consist of several to hundreds of florets (little flowers). Each flower has an inferior ovary, 5 stamens fused at the anthers, and 5 fused petals. The flowers at the center of the head are disk flowers, while the peripheral ones are called ray. Surrounding the head is a series of bracts called the involucre. The calyx, if present, is called the pappus and usually modified into thin hairs for dispersal. It generally crowns the summit of the ovary in the form of awns, capillary bristles, scales or teeth. Nearly all composites are herbs or shrubs. The pollen of many composites is allergenic. The colorful flowers of many species produce yellow and orange dyes.

Yarrow (*Achillea millefolium*)

General Description This is the only species of the genus in Idaho. It is a strongly scented perennial herb with alternate leaves that are finely dissected and appear feathery. The white or sometimes yellow flowers are borne in a flat-topped corymb. Yarrow is widespread and can be found in a variety of habitats from low elevations to above timberline. The generic name honors Achilles. In folklore, his mother dipped the young Achilles into a yarrow bath to make him invincible. Since she held him by his heels, he was made vulnerable through his "Achilles heel."

Field Notes/Uses Yarrow is often referred to as "poor man's pepper." The leaves can be dried, ground, and used as seasoning. The young leaves can be added to salads. The aromatic leaves were also placed in freshly split fish to expedite drying.

Medicinally, the leaves and stems can be dried, boiled in water, strained and drunk to remedy a run-down condition or help with an upset stomach (Willard 1992).

Taken as a hot infusion, yarrow will increase body temperature, open skin pores, and stimulate perspiration, making it a valuable herb for colds and fevers (Moore 1979). The juice can be used as eyewash to reduce redness. Leaves can be used to stop bleeding in small wounds, and to heal rashes when applied directly to the skin. Leaves were also chewed to relieve toothaches (Olsen 1990). A poultice of mashed leaves can be applied to swellings or sores (Train et al. 1957). To date, over 500 biologically active compounds have been identified from the species (Foster and Duke 1990); some are known to be quite toxic. Prolonged use of yarrow may cause allergic rashes and make the skin more sensitive to sunlight.

Rubbing the plant on one's clothing and skin, was an ancient prescription for repelling biting insects. The stalks burned on coals were said to deter mosquitoes. The leaves were used in herbal snuffs and smoking tobaccos. Yarrow has also been used as a hops substitute for brewing Yarrow beer.

American Trailplant (*Adenocaulon bicolor*)

General Description This is a slender, single-stemmed perennial herb up to about three feet tall. The leaves are large, arrow-shaped and arise from near the base of the plant. The disk flowers are white, small and inconspicuous. There is no pappus. The glandular achenes are sticky. The plant is usually found in moist, shady places at low to mid-elevations.

Field Notes/Uses While the edibility of this plant is unknown, the Squaxin used the crushed leaves for use as a poultice (Pojar and MacKinnon 1994, Strike 1994).

False Dandelion (*Agoseris*)

General Description Six species of *Agoseris* occur in Idaho. They are annual or perennial, tap-rooted herbs with milky juice that resemble the common dandelion (*Taraxacum*). The flowers are all ray, yellow or occasionally orange in color. The pappus is white with hairlike bristles. The fruit (achene) is conspicuously ten-nerved. False dandelions occur on moist to dry ground, in meadows and open areas at various elevations. The genus name is from the Greek, meaning "goat chicory."

Field Notes/Uses The leaves and roots of some species are edible when cooked but are bitter, especially in late season (Olsen 1990, Weedon 1996). Strike (1994) indicates that the seeds were eaten by Native Americans in California. The sap from the leaves of some species, when hardened can be used as chewing gum (Pojar and MacKinnon 1994). Since the sap from some species is very thick and insoluble, it may be useful for waterproofing containers (e.g., coiled baskets) and footwear.

Ragweed (*Ambrosia*)

General Description In general, the five species of *Ambrosia* in Idaho are annuals or shrubs with leaves that are opposite below and alternate above. The yellow flowers are arranged in spikes or racemes and the fruit is enclosed in a bur. There is no pappus. Since the wind-blown pollen is highly allergenic, Ragweeds are a notorious cause of hayfever where the plants are common. The genus name is from the Greek and refers to an early name for aromatic plants. It is also the mythic food of the gods.

Field Notes/Uses *Ambrosia trifida* (great ragweed) was cultivated in prehistoric times for its edible seeds in the midwestern United States (Kuhnlein and Turner 1991). Other species were used in teas for various medicinal purposes. The heated leaves of *A. psilostachia* (cuman ragweed) were used as a poultice to ease aching joints, and a decoction was used to bathe bad sores and burns (Strike 1994). Native Americans rubbed the leaves of *A. artemisiifolia* (annual ragweed) on insect bites, infected toes, minor skin eruptions, and hives. A tea was used for fevers, nausea, mucous discharges and intestinal cramping (Foster and Duke 1990). A tea from the leaves of *A. trifida* was formerly used for fevers, diarrhea, dysentery, nosebleeds, and gargled for sore throats (Foster and Duke 1990).

Western Pearly-everlasting (*Anaphalis margaritacea*)

General Description This is a rhizomatous perennial with distinctive white, woolly leaves and stems. The flowering heads are composed of disk flowers with yellow flowers surrounded by conspicuous, papery white involucral bracts. The pappus is comprised of capillary bristles. Pearly-everlasting can be found in various habitats from the foothills to above timberline.

Field Notes/Uses The herbage of western pearly-everlasting has been used as a tobacco substitute to relieve headaches (Weedon 1996). As a tea, the plant has been used for colds, bronchial coughs, and throat infections. The whole plant can be used as a wash or poultice for external wounds. It has also been used for rheumatism, burns, sores, bruises, and swellings (Foster and Duke 1990).

Pussy-toes (*Antennaria*)

General Description In general, pussy-toes are herbaceous often mat-forming perennials. The heads are discoid, with small, white flowers surrounded by bracts that are typically hairy below with a smooth and membranous portion varying in color from white to pink to dark brown or black. The pappus is composed of numerous hairy bristles. The many species (at least 16 species) in Idaho can be found in dry, open habitats, or in moist or seasonally wet places from the foothills to alpine areas.

Field Notes/Uses The sap from the stem of most species can be chewed like gum and has some nutritive value (Kindscher 1987, Weedon 1996, Niehaus 1974). Moore (1979) indicates that a tablespoon of the chopped plant steeped in hot water is an excellent remedy for liver inflammation. It has also been used as an astringent to the intestinal tract (Tilford 1993). Leaves can be poulticed for use on bruises, sprains, and swelling (Foster and Duke 1990). The blossoms could be boiled and used to bathe sore or ulcerated feet, or mashed and applied to sores (Strike 1994). One species, *A. microphylla* (littleleaf pussy-toes) was chewed as a cough remedy by the Thompson Indians in British Columbia. The tiny leaves were also stripped, dried, and used as one of the ingredients in Indian tobacco.

Chamomile (*Anthemis*)

General Description Three species of Anthemis are found in Idaho: *A. arvensis* (corn chamomile), *A. cotula* (stinking chamomile) and *A. tinctoria* (golden chamomile). They are annual or short-lived perennial herbaceous plants with radiate flowering heads composed of white or yellow ray flowers. Introduced European weeds, they are usually found at the lower elevations.

Field Notes/Uses A tea can be made from stinking chamomile to induce sweating and vomiting. An astringent and diuretic, it has been used for ailments such as fevers, colds, diarrhea, dropsy, rheumatism, and headaches (Foster and Duke 1990). The leaves were rubbed on insect bites and stings. Golden chamomile was originally considered a noxious weed of clover fields, but has since been brought into cultivation for horticultural purposes.

Common Burdock (*Arctium minus*)

General Description Common burdock is one of two species found in Idaho. Introduced from Europe, it is a coarse biennial that grows up to five feet tall. The leaves are heart-shaped, with the lower surface slightly hairy, the upper surface glabrous. The heads are discoid with purple flowers. The narrow, hook-tipped involucral bracts spread when in fruit to form the familiar "sticky" burs. Common burdock is a familiar weed of waste places, usually found at the lower elevations. *Arctium lappa* (greater burdock) also occurs in Idaho and can be used in much the same way as common burdock.

Field Notes/Uses Rich in vitamins and iron, the young leaves and shoots can be gathered for use as a potherb, or eaten raw in salad. The plant has a strong rank taste and an objectionable odor. The inner pith-like material of the young stems can be eaten raw, but we find it better when boiled in one or two changes of water. The roots of young plants can be sliced and cooked, then eaten. The older roots can be roasted and ground for use as a tea or coffee substitute. Seeds can be dampened and grown as sprouts (Brown 1983).

The medicinal uses of the plant predate its use as a food plant. The Chinese are said to have used the plants as a blood purifier for thousands of years. Research has confirmed the usefulness of common burdock in the treatment of rheumatism, water retention, and high blood pressure. As a wash, it was used externally for hives, eczema, and skin problems (Foster and Duke 1990). The crushed seeds were used as a poultice.

The tall rigid stems were used as drills for primitive fire starting techniques. The burs can be used as a survival "velcro" for holding clothes together (Brown 1983).

Arnica (*Arnica*)

General Description In general, the many species of arnica are perennials arising from a rhizome or caudex. The leaves are simple and opposite. Flower heads are composed of ray and disk flowers, which are usually yellow or orange. The green involucral bracts occur in one series and the pappus consists of fine white or tawny bristles. Arnicas are usually found in forested or open areas up to the subalpine zone. The name translates as "lamb's skin" and refers to the modified leaves (bracts) that are usually woolly.

Field Notes/Uses All the species are reported to be poisonous if taken internally. Arnica contains arnicin, choline, a volatile oil, arnidendiol, angelic and formic acid, and other unidentified substances that can alter cardiovascular activity. The Federal Drug Administration lists arnica as "unsafe" and bans its use for human consumption. Moore (1979) states that arnica is an external remedy only. The chopped plant is steeped in rubbing alcohol for about a week and squeezed through a cloth. The liniment is then used for joint inflammations, sprains, and sore muscles. It should not be used if the skin is broken since it is toxic if it enters the bloodstream (Willard 1992). Arnica is useful as a topical preparation for bruises, sprains and other closed injuries (Tilford 1993). When gathering, grasp the plant at the base of the stem just below the ground to leave the rhizome for continued growth. Wear gloves as the volatile oils can be absorbed.

Warning All species of arnica are reported to be poisonous if taken internally.

Wilderness Food Storage Pit

To store foods for extended periods of time, Native Americans used storage pits. After digging a hole, moisture was removed from the soil by lining the pit with hot rocks and allowing it to steam. With the rocks left in place, the pit was then lined with dried grasses, and food was placed inside. On top of the food, dried bark from junipers or other plant high in tannic acid were placed to repel insects. On top of this, dried, aromatic, non-poisonous leaves such as sagebrush were placed to disguise the smell of the food. Lastly, the pit was covered with a thick layer of dirt and heavy rocks to prevent

Sagebrush, Wormwood (*Artemisia*)

General Description There is a number of species of *Artemisia* in Idaho, including annual, biennial, and perennial herbs and shrubs. They are mostly aromatic with entire or dissected leaves. The flower heads are small, inconspicuous, and comprised of disk flowers. The genus name honors Artemisia, wife of Mausolus who was the King of Caria (a province in Asia Minor). After the King's death in 350 BC, Artemisia built the renowned Mausoleum, one of the Seven Wonders of the World.

Field Notes/Uses The seeds of many species are edible raw or as flour. The seeds and peeled shoots of *A. douglasiana* (Douglas' sagewort) and *A. ludoviciana* (Louisiana sagewort) were eaten raw by Native Americans in California (Strike 1994).

Herbage of various *Artemisia* species may be toxic if eaten in large amounts, but may be used in small quantities to flavor stews, soups, and other foods. A tea from leaves was a cure for colds and sore eyes, and was used as a hair tonic. Some of the "softer" species can be used as toilet paper and foot deodorant. Crushed leaves can be mixed with stored meat to maintain a good odor. Since many species are aromatic, they can be used to store buried food caches by masking the odor of foodstuff, and to rub on the body to mask human scent while hunting. The wood of *A. tridentata* (big sagebrush) is a good material for fire drills. Although cordage can be made from the bark, it is not very strong.

Many species of *Artemisia* have been used as medicine by Native Americans and were used in sweathouses to relieve numerous ailments. The bitter leaves of *A. absinthium* (absinth sagewort) can be nibbled on to stimulate an appetite (Foster and Duke 1990). A strong tea of *A. ludoviciana* was used as an astringent for eczema, and as a deodorant and anti-perspirant for underarms and feet. A weak tea was used for stomachaches. For sinus ailments, headaches, and nosebleeds, a leaf snuff was used (Foster and Duke 1990). A leaf or root tea of *A. dracunculus* (wormwood) was used for colds, dysentery, headaches, and to promote an appetite. The leaves were poulticed and used for wounds and bruises (Foster and Duke 1990). Moore (1979) says that *A. tridentata* is strongly antimicrobial and was used as an disinfectant and cleansing wash. Volatile oils in *A. tridentata* are responsible for its pungent aroma and are so flammable that they can cause even green plants to burn. It should also be noted that the Federal Drug Administration classifies Artemisia as an unsafe herb containing "...a volatile oil which is an active narcotic poison" (Duke 1985).

Balsamroot (*Balsamorhiza*)

General Description There are four species of balsamroot in Idaho, including *B. hookeri* (Hooker's balsamroot), *B. incana* (hoary balsamroot), *B. macrophylla* (cutleaf balsamroot), and B. sagittata (arrowleaf balsamroot). They are perennial herbs with basal leaves, and thick roots with resinous bark that tastes like turpentine. The heads have large ray flowers that are usually yellow. There is no pappus. Balsamroot can be found in open areas at moderate and low elevations. Balsamroot is often confused with *Wyethia* (mule's ears), which can be found in similar habitats. However, *Wyethia* leaves lack the fuzzy gray appearance seen on the balsamroot.

Field Notes/Uses Although balsamroot is considered one of the most versatile sources of food, it is not necessarily palatable. The plants contain a bitter, strongly pine-scented sap. The large taproot, root crowns, young shoots, young leafstalks and leaves, flower budstalks, and the seeds were all eaten by various Native Americans. The larger mature leaves were often used in food preparation (i.e., wrapping).

The woody taproot of perhaps all species is edible raw or cooked. The polysaccharide inulin is the major carbohydrate found within the root. The roots can be collected throughout the year, but are very difficult to dig out. In some species, the taproot may be as large as one's forearm. Cooking the roots is yet another challenge. One method we have used involves peeling the roots by pounding them to remove the bark. These were then pit cooked for 24 or more hours. When properly cooked, the roots turn brownish and sweet tasting. Another way to prepare the roots is to pit steam large quantities for a day and then mash and shape them into cakes for storage. Cooked this way, the roots were called "pash" or "kayoum."

The young shoots are edible raw or pit cooked before they emerge in early spring. The young stems and leaves can also be eaten raw or boiled as greens. The older stems are fibrous, tough, and will require some additional boiling. The flower budstalks are collected while the buds are still tightly closed, then peeled and eaten raw or cooked as a green vegetable. They have a slightly nutty taste. When harvested from dried heads, the seeds can be roasted and eaten or ground into flour. The chaff is usually removed by winnowing.

The roots are said to be antimicrobial, expectorant, disinfectant, and immuno-stimulant (Tilford 1993, Coffey 1993). They can be mashed and applied to swellings and insect bites. Native Americans considered a boiled solution from the root of *B. hirsuta* (= *B. hookeri* var. *neglecta*) (neglected balsamroot) root to be an excellent medicine for stomach aches and bladder troubles. The mashed roots of arrowleaf balsamroot were also used by Native Americans to treat swellings or insect bites.

Brickellbush (*Brickellia*)

General Description The four species of *Brickellia* that occur in Idaho are perennial herbs with fibrous roots. The disk flowers are all tubular, white or creamy to pink-purple. They can be found in a variety of habitats at the lower elevations. This is a large and complex genus consisting mostly of shrubs.

Field Notes/Uses Moore (1989) states that a tea or tincture from *B. grandiflora* (tasselflower brickellbush) has three distinct uses: 1) lowering blood sugar in certain types of diabetes; 2) stimulating hydrochloric acid secretions by the stomach; and 3) stimulating bile synthesis and gallbladder evacuation. Others species were also probably used medicinally by Native Americans.

Plumeless Thistle (Carduus)

General Description Three species are reported to occur in Idaho: *C. acanthoides* (spiny plumeless thistle), *C. nutans* (nodding plumeless thistle), and *C. pycnocephalus* (Italian plumeless thistle). *Carduus* is distinguished from *Cirsium* in that the pappus of *Carduus* is simple and smooth, not a plume.

Field Notes/Uses Kirk (1975) indicates that the pith of four species (unspecified), without the easily removed rind, may be boiled in salted water and seasoned in various ways. The dried flowers may be used as rennet to curdle milk. Additionally, Strike (1994) indicates that the raw or cooked leaves and stems, and raw buds were also eaten.

Knapweed (*Centaurea*)

General Description The seven species of knapweed in Idaho are considered to be noxious weeds. That is, they are invasive and damaging to native habitats. In some areas of the State, they grow so profusely that they crowd out other species of plants, making the area uninhabitable for native plant and animal species. This genus apparently causes an inability to swallow if ingested by horses, resulting in death.

Field Notes/Uses Historically, knapweed was used as a topical vulnerary, sore throat remedy, and an appetite stimulant. *Centaurea cyanus* (bachelor's button) is considered a powerful nervine, and Native Americans used it for venomous bites, indigestion, jaundice, and eye disorders. Culpeper writes that "Knapweed gently heals up running sores, both cancerous and fistulous, and will do the same for scabs of the head." Though the formulations and preparations used might be considered "questionable," the plants are abundantly available and probably warrant further investigation.

Chaenactis (*Chaenactis*)

General Description These are biennial or perennial herbs from a taproot. The leaves are pinnately dissected and the flowering heads are comprised of disk flowers which are white to pink to rose in color. The five species in Idaho can be found in open, dry and rocky habitats from the lower elevations into the alpine. The genus is endemic to the western United States.

Field Notes/Uses Leaves of *C.*

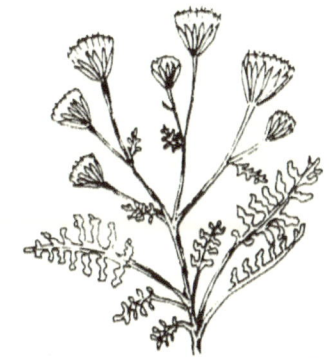

douglasii (Douglas' pincushion) were mashed and used to poultice sprains and swellings. A decoction of the plant was used for indigestion, coughs, and colds. Mashed leaves used on rattlesnake bites (Strike 1994).

Rush Skeletonweed (*Chondrilla juncea*)

General Description This is a perennial with many branched, wiry stems that range from 1 to 4 feet tall. The rosette leaves resemble common dandelion and are hairless with deep, irregular teeth that point back toward the leaf base; they wither by flowering time. The plant has milky juice; coarse, reddish downward-pointing hairs at the base of the single flowering stem; and small yellow flowers and plumed seeds that ride the wind.

Field Notes/Uses It thrives in well-drained sandy or gravelly soils and has invaded extensive areas of shallow silt loam soils in other areas as well. In addition to deep taproots, it has lateral roots that produce daughter rosettes. Plants also grow from buds on root fragments cut by cultivation or other equipment.

Green Rabbitbrush (*Chrysothamnus viscidiflorus*)

General Description The two Greek words *chryso* (gold) and *thamnus* (shrub) describe this genus in bloom, when most species are densely coated with small but distinctly gold or yellow flowers in numerous tiny heads. Each corolla is radially symmetrical, with spreading lobes. Style branches also are elongated and spreading, extending well past the corolla. Fruits generally are long and cylindrical with five ribs and a pappus of many bristles.

Field Notes/Uses The rubber shortage of World War II stimulated research on rabbitbrush and other rubber-producing plants. Rabbitbrush produces a high quality rubber called chrysil that vulcanizes easily. Extraction of this rubber for economic reasons at this point is not feasible. Because of their rubber-based compound, rabbitbrush will burn even if its wet or green. Navajo Indians derived a yellow dye from the flowers, while the inner bark yielded a green dye.

Chicory (*Cichorium intybus*)

General Description This s a perennial herb that grows up to three feet tall with dandelion-like leaves. The blue flower heads, which can be seen from spring to fall, are composed of 15-20 or more ray flowers. The sap is milky. Chicory is a plant of waste places and is found at the lower elevations. Introduced from Europe, it now grows throughout the United States.

Field Notes/Uses While the roasted root was used as coffee, though it is not considered a very satisfactory substitute by itself. Many coffee producers have used chicory as a coffee additive.

The young basal leaves and flowers buds hidden at the base of the leaves are edible and best if collected from fall to spring. Because they are bitter, we found it necessary to boil them in at least 1 to 3 changes of water. When collected very young, the plants are milder when eaten raw. In some European countries, the buds are pickled and canned (Tull 1987).

> **Chicory Coffee Additive**
>
> To make a coffee additive, dig up chicory roots in the fall through spring, scrub them, and slice in half. Roast them in an oven at a low temperature (e.g., 250°F) for 2 to 4 hours, or when they become dark brown and brittle. Break up and grind as you would coffee. One part Chicory to 4 parts coffee is a common ratio when brewing.

Thistle (*Cirsium*)

General Description The many species of thistle that occur in Idaho are characterized as biennial or perennial herbs with alternate leaves that are lobed or cleft with spines. The red, yellow, or white heads are showy and the involucral bracts are overlapping. The native and introduced species can be found in a wide variety of habitats from the foothills to the higher elevations. *Cirsium* comes from the Greek *kirsos*, meaning "swollen vein," for which thistles (kirsios) were a reputed remedy.

Field Notes/Uses Thistles were not a major food source in the past, but were used when needed. Truman Everts, a participant in the early explorations of the Yellowstone Park region, became lost for more than a month and subsisted on thistles. He apparently had lost his glasses and was able to identify thistles by touch. Although Thistles are difficult to collect, they are well worth the pain.

All species have roots that can be eaten raw, boiled or roasted. Some have roots that turn sweet when roasted. The immature flower buds (asparagus-

like) can be eaten raw or cooked. Young leaves de-thorned are edible raw, and a tea can be brewed from all leaves. The peeled young stems may be cooked as greens, and resemble celery in taste. The older stalks are also edible but are somewhat more fibrous and bitter. The seeds can be boiled and eaten in the same manner as sunflower seeds, or they can be ground into flour for baking.

Medicinally, thistle stalks were chewed to ease stomach pains. Pounded stalks were used as a salve for facial sores or on infected wounds. A decoction made from thistle roots was used to relieve asthma (Strike 1994).

When well dried and de-thorned, stems can be used as hand drills for starting fires. The stem fibers of any thistle species can be used as thread or crude cordage. To obtain the fiber, simply soak the stalks in water for a day or more to loosen them from the outer layer. The downy part of seed heads makes good insulating material and a good tinder additive.

Canadian Horseweed (*Conyza canadensis*)

General Description This is an annual weed similar to *Erigeron*, it grows to about two feet tall with numerous, narrow leaves. There are numerous white flower heads. Canadian horseweed is usually found growing in waste places at the lower elevations.

Field Notes/Uses A native to North America, horseweed was introduced into Europe around the mid-17th century where it became widely known for its tonic and astringent properties. A tea was made from the entire dried plant and used for gravel dropsy, diarrhea, and scalding urine. Native Americans used the plant in the form of a tea for leucorrhea, and applied the solution to external sores in cases of gonorrhea (Callegari and Durand 1977). Foster and Duke (1990) also indicate that *Erigeron canadensis* (= *C. canadensis*) was used as a folk diuretic, astringent for diarrhea, kidney stones, nosebleeds, fevers, and cough. The leaves and tops of horseweed can be pounded and eaten uncooked (Strike 1994).

Golden Tickseed (*Coreopsis tinctoria*)

General Description This is an annual that is native to the southern United States and has spread throughout much of North America. The leaves are finely divided and occurring mostly in the lower portion of the plant. The flowers are a vibrant yellow with maroon centers. The genus name comes from the Greek words *koris* meaning bug and *opsis* meaning like in reference to the shape of the seed which resembles a bug or tick. The specific epithet means "used in dyeing." Plants in the genus *Coreopsis* are commonly called tickseed in reference to the resemblance of the seeds to ticks.

Field Notes/Uses The Zuni people use the blossoms of the tinctoria variety to make a mahogany red dye for yarn. This variety was formerly used to make a hot beverage until the introduction of coffee by traders. Women also use an infusion of whole plant of this variety, except for the root if they desire female babies

Hawksbeard (*Crepis*)

General Description Nine species of hawksbeard occur in Idaho. In general, they are perennial, tap-rooted herbs with milky juice. The leaves are alternate or all basal, and the yellow flowers are all ray. The various species can be found in dry open places at lower elevations to gravelly or rocky places in alpine or subalpine areas.

Field Notes/Uses The stems and leaves of *Crepis* were eaten (Strike 1994). The Karok Indians of California peeled the stems of *C. acuminata* (tapertip hawksbeard) before eating (Strike 1994).
The seeds or whole plant of *C. acuminata* was thoroughly crushed and applied as a poultice to breasts after childbirth to induce milk flow (Train et al. 1957). The root of the plant was used to remove a foreign object from the eye. The root can also be ground into a smooth powder and sprinkled in the eye to treat eye problems. Several applications were necessary.

Heathgoldenrod (*Ericameria*)

General Description The genus *Ericameria* is a segregate of the previously recognized genus *Haplopappus* and a few species recently transferred from *Chrysothamnus*. Authorities have separated *Ericameria* from *Haplopappus* and *Chrysothamnus* on the basis that *Ericameria* has phyllaries either equal or unequal in length and these phyllaries are arranged in spirals; whereas in *Chrysothamnus*, the phyllaries are unequal in length and they are arranged in vertical ranks. *Ericameria* is from the Greek *Erica* (Ereika) for heath, and *meris* or *meros* for division or part, referring to the heath-like leaves.

Field Notes/Uses A tea was reported to be made from the twigs of *E. nauseosus* that provided relief from chest pains, coughs, and toothaches. The leaves and stems were also boiled and the liquid was used to wash itchy areas.

Fleabane (*Erigeron*)

General Description Many species of fleabane occur in Idaho. They are characterized as annual, biennial, or perennial herbs with alternate or basal leaves. The flowering heads are radiate with narrow ray flowers that may be white, pink, blue, purple, or occasionally yellow. The numerous disk flowers are yellow, and the pappus is comprised of capillary bristles. The various species bloom mostly in the spring and early summer, except at the higher elevations, and can be found in a variety of habitats. The genus name comes from the Greek *eri* (early) and *geron* (old man). The common name, fleabane, comes from the belief that these plants repelled fleas.

Field Notes/Uses Fleabanes, in general, are listed as astringent and diuretic (Willard 1992). The disk flowers of *E. philadelphicus* (Philadelphia fleabane) were powdered to make a snuff to cause one to sneeze and breakup a cold or catarrh. A tea from the entire plant of *E. annuus* (eastern daisy fleabane) was used to treat a sore mouth. The dried roots, stems, and flowers of *E. peregrinus* (subalpine fleabane) were steeped in hot water and the patient would breathe the vapors. Fleabanes may cause dermatitis in some people.

Woolly Eriophyllum (*Eriophyllum lanatum*)

General Description There is one species in Idaho, *E. lanatum*. It is a hairy perennial with yellow flowers, it is found in dry, open places at lower to mid elevations.

Field Notes/Uses Eriophyllum seeds were parched and ground into flour by Cahuilla and Luiseno Indians in California. The seeds were also incorporated into pinole (Strike 1994).

Spotted Joe-pye Weed (*Eutrochium maculatum*)

General Description The plant 2 to 6 feet in height and the flowers are pink to purple-pink, and arranged in a flat-topped cluster. Leaves are elongate, with dentate outer margins, and arranged in whorls of 4 to 5 leaves. The stem is purple or green spotted with purple.

Field Notes/Uses When in full bloom, the flowering tops can be gathered and stripped from the stalk, then dried and used to make a tea or tonic that causes vomiting (Sweet 1976). The tea has been used as a cold tonic by Native Americans, while a hot infusion is used for malarial fever (Sweet 1976). Spotted Joe-pye weed is thought to have beneficial uses as either a kidney tonic or a urinary tonic. Herbal remedies are only prepared from the root.

Common Gaillardia (*Gaillardia aristata*)

General Description This is a perennial herb with a slender taproot. The leaves are linear to lance-shaped. The flowering heads are solitary or few flowered with yellow ray flowers and purplish disk flowers. It is found in open places at low and middle elevations.

Field Notes/Uses Blackfoot Indians drank a tea made from the root for gastroenteritis, and the chewed, powdered root was applied to skin disorders (Willard 1992). They also bathed sore nipples of nursing mothers in a tea made from the plant or used the liquid as an eyewash or nose drops.

Cudweed (*Gnaphalium*)

General Description Two species of cudweed can be found in Idaho: *G. palustre* (western marsh cudweed) and *G. uliginosum* (marsh cudweed). These are woolly annual, biennial, or short-lived perennials herbs with alternate leaves. The plants are often confused with *Antennaria* , but cudweeds have both male and female flowers on the same plant and are tap-rooted. The disk flowers are yellow or whitish. The species are found from the low to mid elevations in moist, open areas to well-drained soils.

Field Notes/Uses The bruised plant assists in healing wounds, and steeping the leaves in cold water is used for increasing perspiration (Sweet 1976). Some species contain pyrrolizidine alkaloids and should be regarded as potentially toxic (Weedon 1996).

Gumweed (*Grindelia*)

General Description The four species in Idaho are biennial or short-lived perennial of waste places at low elevations. The leaves are alternate and have toothed margins. A sticky, resinous sap covers the leaves and bracts of the yellow flowers.

Field Notes/Uses In general, gumweeds are considered toxic, and the toxicity appears to be dependent upon the soil in which it grows. However, many species have been used medicinally for hundreds of years.

The sticky flowers heads of *G. squarrosa* (curlycup gumweed) were used as a chewing gum substitute. The young leaves make an aromatic, bitter tea. The flower heads can be boiled in water and used as an external remedy for skin diseases, scabs, and sores. A hot poultice of the plant was used for swellings. A tea made from the plant was used for coughs, pneumonia, bronchitis, asthma, and colds.

Broom Snakeweed (*Gutierrezia sarothrae*)

General Description This is a shrub with linear leaves found in dry, open areas at low to mid elevations. Another common name, matchweed, refers to the match-like appearance of the flower heads.

Field Notes/Uses As with many aromatic plants, this species was used medicinally (Hart 1976). The plant was boiled to make a tea for colds, coughs and dizziness. The tops of fresh, mature snakeweed were boiled until strong and dark. The liquid could be drank for lung trouble and colds, or applied externally for skin ailments such as heat rash, poisoning, and athlete's foot. For respiratory ailments, the root was boiled in water and the steam inhaled (Hart 1976).

Common Sneezeweed (*Helenium autumnale*)

General Description This is a plant up to three feet tall with fibrous roots. The leaves are sessile, alternate, and shallowly toothed to entire. The ray flowers are showy, yellow to orange in color, and reflexed. Disk flowers are also yellow, soon turning brown. Common sneezeweed can be found near water at low and mid elevations. Sneezeweed refers to the irritation the pollen causes in sensitive people.

Field Notes/Uses Several species within this genus contain toxic alkaloids (including helenalin) and are considered poisonous to livestock fish, worms, and insects. The roots have been used to relieve rheumatic pains, treat stomach disorders, and cure colic and diarrhea in infants. American Indians used the dried powdered disk flowers of common sneezeweed as a snuff for headcolds and catarrh. A flower tea was used to treat intestinal worms (Foster and Duke 1990).

Sunflower (*Helianthus*)

General Description The seven species of *Helianthus* found in Idaho are annual or perennial herbs. The large ray flowers are yellow; the disk flowers are yellow to red-purple. The seeds appear 4-angled and have two flat awns that when young point upward at the top. Other genera, including *Wyethia*, *Balsamorhiza*, and *Arnica*, are often mistaken for *Helianthus*. The various species can be found in disturbed areas to moist sites in the lower elevations. The genus name comes from the Greek *helios anthes*, which means sunflower.

Field Notes/Uses The largest member of this genus in Idaho, *H. annuus* (common sunflower), is a valuable and useful plant. It has been cultivated in the U.S. since before Columbus. Other species of *Helianthus* may be used similarly.

The seeds may be eaten raw or roasted, then ground into meal and made into bread. The roasted shells can be used as a coffee substitute. To separate large amounts of seeds from shells, first grind them coarsely, and then stir vigorously in water. In this way the shells will float, while the seeds sink to the bottom. The tiny unopened flower buds are also edible with a flavor similar to artichokes. To reduce their bitterness boil in 2-3 changes of water. Serve with lemon and melted butter.

Sunflower oil can be extracted from the seeds for cooking, and can also be used in making soap, paints, varnishes, and candles. It is extracted by simply boiling the crushed seeds and then skimming the oil from the surface of the water (Kirk 1975). The pulp remaining after the oil is extracted also provides food for livestock.

Medicinally, the crushed roots can be applied to bruises (Foster and Duke 1990). Other uses of sunflower include fiber obtained from the stalks for cordage, weaving, and sewing. The Chinese reportedly use the stalk fibers in fabrics and the pulp for paper production; the Russians used the stalks as buoyant material for life preservers (Tull 1987). Purple and black dyes can be obtained from the seeds and a yellow dye from the flowers.

Hawkweed (*Hieracium*)

General Description The nine species of hawkweed in Idaho are fibrous rooted perennial herbs with milky juice. Flowers are all ray, yellow to sometimes orange or white in color. The name "hawkweed" comes from the belief by the ancient Greeks that hawks would tear apart a plant called the *hieracion* (from the Greek *hierax* meaning "hawk") and wet their eyes with the juice to clear their eyesight (Pojar and MacKinnon 1994). Hawkweeds are found in a variety of habitats up to the subalpine.

Field Notes/Uses The green plant and juices of *H. albiflorum* (white hawkweed) may be used as a substitute for chewing gum (Niehaus 1974), although it is best when dried first. The plant was also used to ease toothaches, to cure warts, or as astringent in treating hemorrhages, and as a general tonic (Strike 1994).

Pacific Hulsea (*Hulsea algida*)

General Description This is an attractive yellow composite of high places that grows in crevices on talus slopes at treeline, and flowering toward the end of August. It has serrated leaves, thick stem, and the yellow flowering head is reminiscent at first of a dandelion, but on closer examination one sees that the leaves are thick and sticky, and the flower head has both ray and disk-flowers. The plant, like many of the composites, gives off a pronounced, aromatic odor.

Field Notes/Uses The plant is named for U.S. Army physician and botanist, Dr. Gilbert White Hulse (1807-1883); *algida* means "cold" in Latin, and is a reflection of the plant's environment.

Fineleaf Hymenopappus (*Hymenopappus filifolius*)

General Description Very fine thread-like mounds of leaves are evident for weeks before tall slender stalks arise and are then topped by several small yellow flowers. *Hymenopappus filifolius*, is composed only of disk flowers and is easily overlooked. *Hymenopappus* refers to the membranous pappus (small scales, bristles, or hairs at the apex of the seed) and *filifolius* is from the Latin *fili*, (thread) and *folius* (leaf).

Field Notes/Uses A poultice of the chewed root and applied it to swellings. A decoction of the plants was taken by the Navajo for coughs. The leaves were boiled, rubbed with cornmeal, and baked into bread. The roots were chewed as chewing gum.

Povertyweed (*Iva axillaris*)

General Description This is a long-lived perennial with creeping roots. It is a widespread native that is a desirable component of salt marsh and alkali plains.

Field Notes/Uses In areas that have been disturbed by overgrazing, the plant will form large clonal colonies that once established, will be difficult to eradicate. The pollen of this plant is highly allergenic and plants may cause contact dermatitis in sensitive individuals.

Wild Lettuce (*Lactuca*)

General Description These are tall, annual, biennial, and perennial plants with alternate leaves and milky juice. The yellow, blue, or whitish flowers are all ray. The pappus is white to brownish. The six species in Idaho can be found in moist meadows and disturbed areas at lower elevations.

Field Notes/Uses Collected in the late fall to early spring, the plants should be boiled in a couple of changes of water to reduce the bitterness. The earlier or younger the plant is collected, the better the flavor. Because of the latex sap, raw greens can cause upset stomach if eaten in quantity. In sensitive people, the latex can cause dermatitis (Elias and Dykeman 1982). These wild plants contain more Vitamin A than spinach and a good quantity of Vitamin C

(Tull 1987). An extract of the white sap from two species of *Lactuca* in Europe has been used to replace opium in cough remedies (Tull 1987). The extract, lactucarium, is reported to be a mild sedative (Lewis and Elvin-Lewis 1977). The plants also contain a mildly narcotic compound in the latex. The active constituents increase during flowering and are relatively low in young plants.

Whitedaisy Tidytips (*Layia glandulosa*)

General Description This annual grows up to 24 inches tall and has leaves that are rough hairy, linear to lanceolate in shape, with the basal ones, being toothed or lobed while the upper ones are entire. The flowering heads have both ray and disk flowers present. There are about 25 to 100 disk flowers. Layia is found in sandy soil.

Field Notes/Uses The seeds of *Layia* were often used in making pinole. The seeds of this species are edible after grinding it into flour for mush.

Ox-eye Daisy (*Leucanthemum vulgare*)

General Description The genus name is from the Greek *leukos* (white) and *anthemon* (flower). This is a Eurasian species now naturalized in North America, especially at the lower elevations in parks and lawns.

Field Notes/Uses The species is a popular ornamental, and is used in home remedies for catarrh and is more or less edible. An infusion of flowers and roots was used as eyewash, and the flowers were used to make a tonic.

Skeletonplant (*Lygodesmia*)

General Description There are two species of *Lygodesmia*: *L. grandiflora* (largeflower skeletonplant) and *L. juncea* (skeletonplant), in Idaho. They are annual or perennial herbs with a rush-like appearance and milky juice. The flowering head is comprised of all ray flowers that are pink or purple with 4-8 involucral bracts. In general, the species are found in found in dry, open places at lower elevations.

Field Notes/Uses Both species were reportedly drunk as a tea by Native American women to increase their milk flow. The bluish colored tea of *L. juncea* was also believed to give the mother an "inner power" which was then passed on to the infant. The sap of *L. juncea* can be chewed like gum. A tea made from the whole plant is said to cure diarrhea (Hart 1976).

The leaves of *L. grandiflora* were boiled with meat or mush. An infusion of the plant was used to treat sore eyes (Willard 1992).

Tarweed (*Madia*)

General Description Four species of tarweed occur in Idaho. Typically, they are annuals with a tar scent of varying intensity. The leaves are narrow, usually opposite below and alternate above. The flower heads are comprised of inconspicuous yellow ray flowers. Tarweed can be found at moderate elevations in open, grassy, or vernally moist areas.

Field Notes/Uses The seeds of *M. glomerata* (mountain tarweed) may be eaten raw, cooked or dried and ground into meal. The scalded seeds also yield nutritious oil. All tarweeds were used medicinally by old Spanish settlers (Parsons 1966). An oil of excellent quality was made from their seeds in this country before olives were readily available.

Strike (1994) indicates that *Madia* seeds were collected and stored until needed. Seeds were often used in making pinole by many Native Americans in the west. In California, Native Americans pulverized *Madia* seeds and ate them dry. When *Madia* seeds had matured but the plants were still green, Hupa Indians burned the areas were the plants grew. The seeds were then gathered from the scorched plants and because they needed no further parching, they were crushed into flour. The roots of some species were also eaten.

Sowthistle Desertdandelion (*Malacothrix sonchoides*)

General Description The short, lobed, basal leaves are diagnostic even before the flower emerges. The plant is typically between 2 and 14 inches tall, often with outward leaning flower stems, and flower heads are composed only of ray flowers. Two other species occur in Idaho: *M. glabrata* (smooth desertdandelion) and *M. torreyi* (Torrey's desertdandelion).

Field Notes/Uses The two Greek words "*malacos*" and "*trichos*," mean "soft hairs" and refer to the downy hairs of the pappus (hair-like structures at the top of Asteraceae growing seeds). The flower of desertdandelion looks like the flower of sow thistle (*Sonchus oleraceus*), thus the species name, *sonchoides*. Several Native American tribes ate the seeds of a related species (*M. californica*).

Disc Mayweed (*Matricaria discoidea*)

General Description These are annual or biennial herbs that have a branched habit. Leaves are alternate and pinnately lobed or divided. The small, terminally arranged flower heads are composed of disk or ray flowers. *Matricaria* are introduced plants that are circumboreal in distribution. The scientific name is from the Latin *mater* or *matrix*, referring to the plants reputed medicinal value.

Field Notes/Uses Another common name for this species is pineapple weed. Crush the plant and smell your fingers. A delicious tea can be made from the dried flowers of the plant. The leaves are edible, but bitter. The medicinal uses of pineapple weed are identical to that of chamomile (*Anthemis*). Used as a tea it is a carminative, antispasmodic, and mild sedative.

Nodding Silverpuff (*Microseris nutans*)

General Description This is a taprooted perennial with milky juice. The leaves are basal, and the flower heads are always ligulate and yellow. Look for the plant in open and moist habitats of the montane and subalpine zones.

Field Notes/Uses The slender roots of nodding silverpuff are apparently edible raw (Weedon 1996).

Ragwort (*Packera*)

General Description Members of this genus were previously included in the genus *Senecio*. They are mostly perennial herbs with basal or alternate entire to deeply lobed leaves. The involucral bracts are uniseriate. Ray flowers are yellow or orange or lacking, and the pappus consists of capillary bristles. This is a complex genus with some species, especially *Packera paupercula*, being highly variable. You should include vegetative basal rosette leaves (or take notes about their shape).

Field Notes/Uses Like *Senecio*, many contain highly toxic alkaloids and should therefore be avoided.

Arctic Sweet Coltsfoot (*Petasites frigidus*)

General Description This is a perennial with creeping rhizomes and thick, succulent, densely hairy stems. Its arrowhead shaped leaves are large, basal, and toothed. The flowering heads are composed of white or purplish ray and disk flowers. The plant can be found in wet habitats at low elevations.

Field Notes/Uses The flowering stems can be boiled as a potherb. The young leaves are edible, but because of their felt-like texture, they are not very pleasant. The aboveground parts have a mild, distinctive salty taste and could be used as a salt substitute. A salt substitute was also made by burning the leaves of a related species, *P. speciosa*, and using the ashes.

The species is famed for its cough relieving abilities. Syrups are recommended for coughs, bronchial congestion, and shortness of breath. The plant contains petasin, a highly effective antispasmodic. As a decoction, it can be used to wash skin eruptions. A poultice made from the leaves was used for sores, insect bites, swelling, and pain. Use the plant only in moderation as it contains alkaloids that can irritate the liver in large quantities.

Rattlesnake-root (*Prenanthes*)

General Description Two species, *P. alata* (western rattlesnake-root) and *P. sagittata* (arrowleaf rattlesnake-root) occur in Idaho. They are perennial plants with milky juice. The leaves are alternate, thin, and the lower ones have wing-margined petioles and an arrowhead shape. The flowering head are nodding and composed of white ray flowers. The species are common on shaded streambanks and other moist substrates in the mountains.

Field Notes/Uses While no records of edible or medicinal for the Idaho species have been found, a root tea from a related species (P. alba), was used as a wash for "weakness." The leaves were poulticed and used on dog and snakebites.

Upright Prairie Coneflower (*Ratibida columnifera*)

General Description Prairie coneflower is a perennial about a foot and a half tall. Several stems usually grow from the crown of a taproot. Leaves about three inches long are divided into five to nine narrow leaflets. The upper one-third of the stem is bare except for the flower heads. The flower heads consist of several hundred tiny purplish-brown flowers that form a cylinder about one inch long. At the bottom of the cylinder appear about a half-dozen bright yellow rays about an inch long. Fruits are tiny, winged achenes about 1/16-inch long. The plant is sometimes grown as an ornamental.

Field Notes/Uses The genus *Ratibida* was named by wanderer-botanist named Constantine Rafinesque-Schmaltz (1773-1840), who often assigned unexplained names to plants. The specific epithet "*columnifera*" is Latin meaning "bearing columns" in reference to the long cylindrical flower heads.

Prairie coneflower is consumed by elk, mule deer, white-tailed deer and pronghorn antelope. Seeds of prairie coneflower provide fair feed for upland game birds, small non-game birds, and small mammals. Native Americans made tea from the leaves and dye from the flowers. Cheyenne Indians boiled prairie coneflower leaves and stems to make a solution to apply externally to draw poison from rattlesnake bites and to provide relief from poison-ivy.

Coneflowers (*Rudbeckia*)

General Description Coneflowers are tall biennials or perennials with alternate leaves. The large flower heads have a cone-like appearance. The three species in Idaho can be found in moist, open, or partly shaded habitats in the montane and subalpine elevations.

Field Notes/Uses Several species are suspected of poisoning livestock when eaten in quantity (Weedon 1996). A root tea from *R. hirta* (black-eyed Susan) was used for worms and colds, and as an external wash for sores, snakebites, and swelling. The root juice can also be used for earaches (Foster and Duke 1990). A root tea from *R. lacinata* (cutleaf coneflower) was drunk for indigestion, a poultice made from the flowers was applied to burns, and the cooked spring greens were eaten for "good health" (Foster and Duke 1990).

American Saw-wort (*Saussurea americana*)

General Description American saw-wort is a perennial that arises from a woody rootstock. The leaves are lance-shaped coarsely toothed and have a broad petiole. The purple flowering heads occur in a dense cluster subtended by leaves at the top of the stem. The broad involucral bracts of the heads have rounded tips, and all of the flowers are disk flowers (no ray flowers). The achene has a bristly pappus on top; the longest bristles united at the base.

Field Notes/Uses The genus name is for Theodore (the son) and Horace Benedict de Saussure (the father; 1740-1799), two eminent Swiss naturalists. Horace was a geologist, botanist, and early mountain climber. The genus contains about 300 species of perennial herbs with simple to pinnately lobed leaves and flowers in daisy-like heads that lack distinct rays. Some species are grown as ornamentals.

Ragwort, Groundsel (*Senecio*)

General Description Many species of ragwort occur in Idaho. They are annual, biennial, or perennial herbs with alternate or basal leaves. The flower heads are yellow and the pappus is made up of hair-like bristles. Ragworts can be found in various habitats and elevations. *Senecio* is one of the largest genera of plants with nearly 2,000-3,000 species distributed worldwide. Approximately 100 species are found in the Western United States.

Field Notes/Uses Many contain highly toxic alkaloids and should therefore be avoided. A related species, *S. douglasii* (Douglas'groundsel), was used medicinally by southwestern Native Americans as a laxative (Spellenberg 1979, Strike 1994), although misuse could result in death. Strike (1994) indicates that young *Senecio* leaves were eaten by Maidu Indians in California as cooking herbs. Additionally, the seeds may have been eaten by the Chumash Indians (California). *Senecio* leaves were apparently used to line earth ovens.

Goldenrod (*Solidago*)

General Description The eight species of goldenrod in Idaho are perennial herbs with fibrous roots. The leaves are alternate, simple, and either tooth or entire. The heads are made up of yellow ray flowers. Goldenrods may be found in dry to moist habitats from the foothills to timberline, often in dense patches. *Solidago* means "to make whole."

Field Notes/Uses Young leaves can be prepared as potherbs or added to soups. Depending on habitat, age, and personal preference, their palatability is quite variable. The dried leaves and dried, fully expanded flowers can be used to make a tea. The seeds can be used to thicken stews (Olsen 1990). Large amounts of the raw herbage should be avoided as it may be toxic.

Medicinally, *Solidago* was employed for checking internal and external bleeding. An antiseptic lotion may be made by boiling the stems and leaves, or by using dry, powdered leaves (Tilford 1993, Willard 1992, Sweet 1976). The powdered dry leaves were also sprinkled on cuts as a styptic. For insect bites and minor scrapes, apply fresh, crushed or chewed leaves. A tea wash is said to be good for rheumatism, neuralgia, and headaches (Willard 1992).

The fluffy down from the flower heads are a good additive for tinder bundles. All goldenrods contain small quantities of natural rubber, and were once cultivated as a domestic source.

Sow Thistle (*Sonchus*)

General Description The three species of sow thistle that may be encountered in Idaho include *S. arvensis* (field sow thistle), *S. asper* (spiny sow thistle), and *S. oleraceus* (common sow thistle). Introduced from Europe, they are weedy perennials and annuals with alternate leaves that are entire to pinnately divided. The leaf bases have eared-shaped lobes and the margins are prickly. The flower heads are composed of entirely yellow ray flowers. The pappus is bristly. The various species can be found in waste places in lower elevations. The common name is said to be derived from the observation that pigs eagerly consumed the plants.

Field Notes/Uses The young plants of all three species can be prepared as a potherb. As they get older they become increasingly bitter. We found that boiling them in at least two changes of water makes them a little more palatable. Since the plants have an abundance of soluble vitamins and minerals, use only a minimum amount of water and boil briefly. The milky gum obtained from *S. oleraceus* was once used in treating opium addiction, and Native Americans used a tea made from the leaves of *S. arvensis* to calm nerves (Foster and Duke 1990). In Europe, a poultice from the leaves was used as an anti-inflammatory.

Wirelettuce, Skeleton Weed (*Stephanomeria*)

General Description These are more or less branched annual or perennial herbs with milky juice. The leaves are small and often scale-like. The flowers are pink and composed of ray flowers. The four species in Idaho are found in dry, open places at low and mid elevations.

Field Notes/Uses Although there are no documented uses of Idaho species, *Stephanomeria virgata* exudes a milky sap and was used as an eye medication by the Kawaiisu Indians in California (Strike 1994). The sap of another related species (*S. pauciflora*) was used as a chewing gum.

Common Tansy (*Tanacetum vulgare*)

General Description This plant grows up to 60 inches tall and emanates a peculiar spicy odor. The leaves are alternate, deep green in color, deeply cleft, and deeply cut again to give a fine-toothed appearance. The flower heads are small, bright yellow, and comprised of disk flowers only. Two other species of *Tanacetum* are reported to occur in Idaho.

Field Notes/Uses Common tansy, also known as golden buttons and garden tansy, is a perennial herb native to Europe, and has a long history of medicinal use. It was first introduced to North America for use in folk remedies and as an ornamental plant. It is considered to be a noxious weed in many areas.

The first historical records of common tansy cultivation are from the ancient Greeks who used it for a variety of ailments. It was grown in the garden of Charlemagne the Great in the eighth century and in the herb gardens of Swiss Benedictine monks as a treatment for intestinal worms, rheumatism, fevers and digestive problems. Common tansy is still used in some medicines and is listed in the United States Pharmacopoeia as a treatment for colds and fever.

Warning The plants contain alkaloids that are toxic to both humans and livestock if consumed in large quantities. Cases of livestock poisoning are rare, though, because tansy is unpalatable to grazing animals. Human consumption of common tansy has been practiced for centuries with few ill effects, yet the toxic properties of the plants are cumulative and long term consumption of large quantities has caused convulsions and even death. In addition, hand pulling of common tansy has been reported to cause illness, suggesting toxins may be absorbed through unprotected skin.

Dandelion (*Taraxacum*)

General Description Dandelions need very little introduction. All four Idaho species are tap-rooted perennials with milky juice and leaves that form a dense, basal rosette. The solitary flower head is composed of bright yellow ray flowers. They are found in a variety of habitats up to the alpine zone. *Taraxos* means "disorder," and *akos* means "remedy." The plant was an ancient remedy for ailments ranging from spring doldrums to mononucleosis. The latex-like sap was a folk medicine for warts.

Field Notes/Uses *Taraxacum officinale* (common dandelion) was introduced into North America by European settlers as a food crop and a

medicinal cure-all. Every part of common dandelion is edible. The young leaves may be eaten raw or cooked like spinach. The older leaves are also edible, but we find it is better to boil the older leaves in 1 or 2 changes of water to eliminate the bitterness that comes with age. The plants are high in Vitamins A and C, a good source of B complex, and iron, calcium, phosphorous, and potassium. The roots can also be eaten raw, or boiled as a vegetable, baked as potatoes, or added to soups and stews. The roasted root can be used as a substitute for coffee, but it lacks the caffeine buzz. The flower buds can be pickled and added to meals such as omelettes. In general, common dandelion is good for blood circulation.

Taraxacum laevigatum (rock dandelion) is similar in appearance to the common dandelion. It may be found in similar habitats and used in the same way (Harrington 1967). During World War II, a species of dandelion was cultivated by Russians as a commercial source of rubber. In spite of its many uses, many Americans consider the dandelion a pest and spend much time and money trying to eradicate it.

Horsebrush (*Tetradymia*)

General Description Horsebrush is a rather low growing, multi-branched unarmed or spiny shrub, found either as well-scattered individuals or as small colonies mixed in with other vegetation. Some species may reach heights of 6-8 feet but are more commonly 3 feet or less. Horsebrush is commonly associated with the sagebrush vegetation type, but the genus has widespread occurrence from barren slopes and alkaline plains upward into the pinyon-juniper and yellow pine types. "*Tetradymos*" means four-sided, referring to the shape of the flower bracts. The plants are easily mistaken for rabbitbrush (*Chrysothamnus*).

Field Notes/Uses Horsebrush provides ground cover and soil stability. It is generally considered of low forage value, although buds and new leaders are consumed by cattle, sheep, goats, antelope (*Antilocapra americana*), and mule deer (*Odocoileus hemionus*) (McArthur and others 1979). Most species are poisonous to sheep, especially smooth horsebrush (Johnson 1974; Kingsbury 1964).

Flowers are used by small moths, bees, flies, and beetles (McArthur and others 1979). Gelechiid moths form galls in leaves and stems (Hartman 1984).

Sheep that feed on horsebrush just following or in conjunction with black sage (*Artemisia nova*) and then are exposed to bright sunlight may develop a characteristic swelling of the head, called bighead. Sheep grazing the horsebrush may die without developing bighead. The toxins that have been identified in horsebrush are resins and furanoeremophilanes (furanosesquiterpenes) such as tetradymol. Sheep vary considerably in their susceptibility to horsebrush, but often larger amounts may result in death. The horsebrush toxins synergize with other sagebrush toxins, especially black sagebrush.

Townsend Daisy (*Townsendia*)

General Description These are taprooted annual, biennial, or perennial herbs with alternate or all basal leaves. The flowering heads are solitary, few, or many-flowered. The ray flowers are blue or purple to pinkish or white. The achenes are flattened, 2-nerved and usually pubescent. There are eight species in Idaho.

Field Notes/Uses There are about 20 species in this western North American genus. The genus is distinguished from asters by the pappus which consists of several or many awns or scales, or both, instead of hair-like bristles. The genus name honors David Townsend (1787-1858) an amateur botanist from Pennsylvania.

Salsify, Goat's Beard (*Tragopogon*)

General Description The four species in Idaho are introduced, tap-rooted, biennial herbs with milky juice. The leaves are alternate, entire, sessile and clasping at the base and taper to a long point. The flower heads are solitary and composed of pale yellow or purple ray flowers. The heads open early in the day, close about noon and remain closed on cloudy, rainy days. They are found in many habitats at lower elevations. The genus name is Greek, for "goat's beard," probably referring to the thin, tapering, tufted, grass-like leaves.

Field Notes/Uses The fleshy roots of the purple flowered *T. porrifolius* (salsify) can be eaten raw or after cooking. The flavor resembles that of an oyster, an acquired taste! The yellow flowered species, *T. dubius* (yellow salsify) and *T. pratensis* (meadow salsify), are also edible, but are somewhat smaller, more fibrous, and tough. Salsify root has been cultivated for over 2,000 years in the Mediterranean. The young leaves and stems of all species can be eaten after boiling until tender. The coagulated sap can be used as chewing gum and as a remedy for indigestion.

Mule's-ears (*Wyethia*)

General Description Three species occur in Idaho. They are stout perennial, simple stemmed herbs with large, erect, alternate leaves. The ray flowers are long and yellow. Heads are usually solitary. All species have leaves on the stems distinguishing them from *Balsamorhiza*, which only has leaves at the base.

Field Notes/Uses The seeds are edible and somewhat resemble sunflower seeds in taste (Coffey 1993, Olsen 1990). The roots of *W. helianthoides* (sunflower mule-ears) can be eaten after they have been cooked for a day or two, in a steam pit (Sweet 1976). Regardless of how long the roots are cooked, the smell is almost intolerable to enjoy eating. A decoction of leaves was used as a bath, producing profuse sweating. The leaves are considered to be poisonous and should not be taken internally. The Klamath Indians used the mashed root as a poultice for swellings (Sweet 1976).

Cocklebur (*Xanthium*)

General Description Two species, X. spinosum (spiny cocklebur) and *X. strumarium* (rough cocklebur) occur in Idaho. These are coarse annual weeds of uncertain origin that have a cosmopolitan distribution. The stems of the plants are simple, and the leaves are alternate. The flower heads are solitary or clustered in the leaf axils. The bur (seed) has conspicuous, slender hooked prickles. Cockleburs can be found in low elevation floodplains.

Field Notes/Uses The uses of cocklebur are primarily medicinal. They have been used by many aboriginal people throughout North and South America, and as a herbal medicine in China. The seeds were ground, mixed with corn meal, made into cakes or balls, and steamed by the poorer class of Zuni Pueblo (Dunmire and Tierney 1995).

Historically, the roots of *X. strumarium* were used for scrofulous tumors (related to the lymph glands in the neck). Both species were also used for rabies, fevers, and malaria. They possess diuretic, fever reducing, and sedative properties (Foster and Duke 1990). Native Americans used a leaf tea for kidney disease, rheumatism, tuberculosis, diarrhea, and as a blood tonic. The seeds have germicidal qualities, and were ground and applied to wounds. The seeds also contain oil that can be used as lamp fuel.

Touch-me-not Family - BALSAMINACEAE

These are succulent herbs with swollen nodes and dentate, alternate leaves. The five-petalled flowers are irregular with three sepals, the lowermost forming a tubular spur. The fruit is an explosive capsule. Four genera and 500-600 species are found primarily in Eurasia, North America, and Africa. The genus Impatiens is native to the eastern United States. The family is of no direct economic importance.

Touch-me-not (*Impatiens*)

General Description Four species are found in Idaho. These are annual plants with somewhat ovate leaves, and are found in moist soils in forests or along streams from low to mid elevations.

Field Notes/Uses The young shoots of *I. capensis* can be used as potherbs after boiling in two changes of water. If you collect the plants in the spring when they are less than six inches tall, they are tender and require less cooking since the plants contain calcium oxalate crystals, frequent use is discouraged. The ripened seeds are edible raw, and can be used as a dessert topping.

Best known as a field remedy for poison ivy and stinging nettle, *I. capensis* is found in the same habitat as poison ivy. The mucilaginous juice from the stems (before flowering) was applied to the rash (Foster and Duke 1990). As a folk remedy, the poultice was used for bruises, burns, cuts, insect bites, sores, and sprains (Foster and Duke 1990). The extract spoils quickly, but retains medicinal properties longer when frozen. The plant also contains an antifungal agent (2-methoxy-1, 4-napthoquinone) that provides a useful and effective treatment for athlete's foot (Elias and Dykeman 1982).

Barberry Family - BERBERDIACEAE

There are nine genera and 590 species in this family, distributed throughout the northern hemisphere and South America. Two genera are native to the United States. This is a diverse family of perennial herbs and shrubs. The flowers have six or more stamens that split open by two hinged valves to splatter pollen over insects as they crawl by. Several species are cultivated as ornamentals.

Oregon-grape, Barberry (*Berberis/Mahonia*)

General Description The four species of barberry in Idaho are evergreen shrubs with yellow wood and flowers. The leaves are alternate, simple

or compound often with spiny margins. The succulent fruits are purplish-black or red berries. Barberries can be found in moist or dry woods and hills at moderate elevations. Members of the genus *Mahonia* are treated here as *Berberis*.

Field Notes/Uses The blue berries are edible raw or can be dried for winter use or added to soups to improve flavor. We like them best when picked right off the plant. They also make good jellies.

Medicinally, the bark may be boiled and the infusion used to wash sores on the skin and in the mouth (Willard 1992, Moore 1979, Sweet 1976). The plants contain berberine, a bitter alkaloid that gives roots their distinctive yellow color and usefulness as a digestive tonic. Berberine stimulates the involuntary muscles (Hart 1996) and possesses anti-pyretic, laxative, and anti-bacterial qualities (Tilford 1993). A tea from the berries of B. vulgaris (common barberry) can be drunk to stimulate an appetite (Foster and Duke 1990). The liquid obtained from the root by chewing was placed on injuries and wounds (Willard 1992)

A yellow dye can be obtained by boiling bark and roots. The whitish film on the berries is yeast that can be used in making a primitive sourdough starter.

Birch Family - BETULACEAE

Of the six genera and 150 species in the Birch Family, five genera are native to the United States. Members of this family are trees and shrubs with deciduous, simple, alternately arranged leaves that have toothed margins. The unisexual minute flowers are arranged in catkins. Both male and female flowers are borne on the same plant. The male catkins are soft and pendant. After releasing abundant pale yellow pollen in early spring, they drop off. The hard female catkin is either erect or pendant, and appears cone-like. Most members of this family grow in moist soil, particularly along streams. Economic products of this family include lumber, edible seeds, and oil of wintergreen.

Alder (*Alnus*)

General Description The four species of *Alnus* in Idaho are small trees or shrubs with smooth, reddish or gray-brown bark. Leaves are egg-shaped and have serrate edges. The male catkins are grouped near the end of branches and drop off after pollen is shed. The female catkin is cone-like and persistent. These plants are usually associated with riparian and wetland sites at low to mid elevations.

Field Notes/Uses The edible catkins are high in protein, but generally don't taste very good. The catkins are more tolerable if they are nibbled raw, added to soups, or dried and powdered and used as a spice. The inner bark is palatable only for a short time in the spring when it is less bitter. A patch of bark is removed from the tree and the tissue scraped off and eaten fresh or dried in cakes.

The bitter leaves and inner bark act on the mucous membranes of the mouth and stomach to stimulate digestion. A tea made from the leaves was used as a wash or a soothing remedy for poison ivy, insect bites, and other skin irritations. Used fresh, the inner bark is emetic, taken to induce vomiting if poisonous substances are ingested. A decoction from alder bark was used to treat colds and stomach trouble. The decoction was also used to reduce pain from burns and scalds (Willard 1992). Chewing alder bark is said to turn one's saliva red, which was used to dye basketry material.

Alder is valued for its hardwood and is useful for open fires as it does not readily spark. It is used widely by aboriginal peoples for woodworking, including dishes, spoons, and platters. The wood is also used for making fire drill sets. The astringent bark and woody cones are used for tanning leather. A black-brown dye from the bark was used for coloring fishing nets to make them more invisible (Tull 1987). Since alder usually grows in the vicinity of free flowing water, it is considered a botanical indicator of water. The roots have small nitrogen nodules that improve the soil for other plants. Alders are good for controlling erosion and floods, and for stabilizing streambanks.

Coal Burning

Have you ever tried to carve a depression in wood to make a spoon, cup, or bowl with only a knife? It can be frustrating at times

To simplify the task, try using hot coals. Place a piece of coal where you want the wood to be hollowed-out, and then blow on the embers with a steady, thin stream of air to keep the coal glowing. A straw is often helpful in directing the air stream. After the coals have burned down a bit, scrape out the charcoal. Repeat the process with fresh coals until the depression is formed.

Birch (*Betula*)

General Description The five species found in Idaho are deciduous trees or shrubs with simple, alternate, and sharply toothed leaves. Birches can be found along streams, and in wet meadows and bogs from the foothills to upper montane zone.

Field Notes/Uses Young birch leaves can be added to salads. The inner bark can be dried and ground into flour, and the twigs can be steeped in hot water for a tea (Peterson 1977). The juice of birch leaves makes a good mouthwash (Weedon 1996, Brown 1983). A tea made from the leaves of B. papyrifera (paper birch) and a poultice of the boiled bark was used to treat bruises, wounds, and burns (Willard 1992). Birch contains a significant amount of methyl salicylate and is often used in teas for headaches and rheumatic pain. Birch is highly regarded as a medicinal plant in Russia and Siberia for treating arthritis.

Since it burns even when wet, birch bark make good tinder. Some Native Americans took the new soft wood, chopped it very fine and mixed it with tobacco. The sap, collected in much the same manner as maple, was sometimes made into vinegar. The best time for tapping is early spring, before the leaves unfurl. The hard wood of some species is used for veneers, cabinet work and interior finish. It is also used for paper pulp, wooden-ware and novelties. Twigs of some species are distilled to produce oil of wintergreen.

Caution Oil of wintergreen, used externally for pain (i.e., muscular, joint, arthritic and rheumatic), may cause irritation to the skin. As a precaution, test by using a few drops of wintergreen oil in a carrier oil or liniment.

Beaked Hazelnut (*Corylus cornuta*)

General Description This is a shrub growing up to 10 feet tall with thin, downy, double-toothed leaves. The nuts are wrapped in an "envelope" of leafy bracts that extend to form a fringed tube longer than the nut. This is a widespread species found at the lower elevations on well-drained soils. The genus name is Latin for "hazelnut."

Field Notes/Uses The sweet nuts of beaked hazelnut ripen in late summer and can be ground into meal and made into bread. The best time to collect the nuts is in early autumn. They may be stored until ripe, and then eaten raw or roasted. The long flexible shoots can be twisted into crude cordage or used as splints in basketry. The roots are said to produce a blue dye, but the method used to obtain the color is unknown (Strike 1994).

Borage Family - BORAGINACEAE

The Borage Family has approximately 100 genera and 2,000 species, with 22 genera native to the United States. Members of the Boraginaceae can be identified by their alternate leaves, round stems, coiled racemes, and 5-parted radially symmetrical flowers. The corolla has a narrow tube that is abruptly flared at the top. The fruit, comprised of 4 nutlets, is helpful to correctly identify species in this family. The name borage comes from a Middle Latin source, *burra*, meaning rough hair or short wool as many of the plants in the family are covered with stiff hairs.

Fiddleneck (*Amsinckia*)

General Description Fiddlenecks are coarse annual herbs with stiff hairs. The flowers are in a scorpion tail-like spike. The three species of *Amsinckia* in Idaho are found on dry soils at low elevations.

Field Notes/Uses The seeds of *A. menziesii* (Menzies' fiddleneck) were pounded into flour, then made into cakes and eaten without cooking (Strike 1994). Coastal California tribes also ate the young fiddleneck leaves. A related species (*A. douglasiana*) was apparently used medicinally (Strike 1994).

Catseye, Popcorn Flower (*Cryptantha*)

General Description The many species of *Cryptantha* in Idaho are annual or perennial herbs that are rough to the touch. They have linear or spatulate leaves and the inflorescence is scorpion tail-like with small, white or yellow flowers. Popcorn flowers are usually found in dry, open areas at various elevations.

Field Notes/Uses While there are no recorded uses of Idaho species, Strike (1994) indicates that the seeds of some *Cryptantha* may have been eaten by the Native Americans in California.

Gypsyflower, Houndstongue (*Cynoglossum officinale*)

General Description This is a tap-rooted perennial herb up to three feet tall. The flowers are dull reddish-purple, and the nutlets are densely covered with short, hooked prickles. This introduced Eurasian weed is common in disturbed areas, especially along logging roads and heavily used pastures.

Field Notes/Uses The plant contains cynoglossine, consolin, and allantoin, a highly effective healing substance that was sometimes used to relieve pain. The use and effect of houndstongue is similar to comfrey in treating skin and intestinal ulcerations (Moore 1979). The root and leaves of the plant make an effective tea for a sore throat accompanied by dry hot cough (Tilford 1993). A leaf poultice was used for insect bites and piles, and other minor injuries such as bruises and burns (Foster and Duke 1990).

Caution Though the young leaves of this plant are edible in small amounts after boiling, internal use of houndstongue is not recommended. The plant contains potentially carcinogenic alkaloids that may be harmful to the liver if ingested in large quantities.

Arctic Alpine Forget-me-not (*Eritrichium nanum*)

General Description These are dwarf, cushion-like perennial herbs. The leaves are densely crowded on numerous short shoots, and more or less hairy. The blue flowers with a yellow center occur in clusters. There are 1-4 nutlets.

Field Notes/Uses The showy blue flowers of arctic alpine forget-me-not are pollinated by a wide array of insects ranging from syrphids to bees. Seeing these beautiful, blue alpine plants is a special treat for hikers and climbers. It is the just award for reaching the "top of the mountain."

Stickseed (*Hackelia*)

General Description Ten species of *Hackelia* occur in Idaho. These are mostly tall, taprooted perennials with numerous blue or white flowers with yellow centers. Each of the flowers gives rise to four small nutlets that possess rows of barbed prickles down their edges - hence the name stickseed. It is these prickles that readily enter clothing or the fur of animals, but retard their being pulled out again. The seeds are then transported and the plants become established a long distance from where they originated. The various species grow in moist to medium dry soils in the foothills to above 8,000 feet. The genus is named for Joseph Hackel, a Czech botanist who lived from 1783 to 1869.

Field Notes/Uses No reports of any species being eaten by humans. However, the majority of the species are reported to be "noxious," in that, their fruits tend to cling in great numbers to wool shirts, socks, and trousers. Additionally, *H. floribunda* (manyflower stickseed) and *H. hispida* (showy stickseed) were reported to be used as medicinal plants by several Native American tribes. The prickles on the fruit of manyflower stickseed can also cause skin irritation and swelling.

Salt Heliotrope (*Heliotropium curassavicum*)

General Description Flowers small, blue or white, with 5 petals fused to form a tubular corolla with free lobes at the apex. Flowers arranged in strongly curled sprays. Stem and leaves smooth, fleshy. Plant found in open mats along the shore. Leaves elongate, alternate, pointed at the tip, and with smooth outer margins.

Field Notes/Uses The genus name is from the Greek, *helios*, meaning sun, and *tropos*, for turning. In some parts of the West this species is called Quail Plant after the birds that feed on its fruit, and the Spanish name, Cola de Mico, meaning "monkey tail," describes the coiled flower cluster.

The ashes of the burnt plant are used for salt. Consumption of a tea made from the leaves of this plant can cause liver disease. Sensitivity to a toxin varies with a person's age, weight, physical condition, and individual susceptibility. Children are most vulnerable because of their curiosity and small size. Toxicity can vary in a plant according to season, the plant's different parts, and its stage of growth; and plants can absorb toxic substances, such as herbicides, pesticides, and pollutants from the water, air, and soil.

Stickseed (*Lappula*)

General Description Two species are listed as occurring in Idaho - *L. occidentalis* (flatspine stickseed) *and L. squarrosa* (European stickseed) are non-native, introduced species. They are taprooted annual herbs with linear, alternate leaves. The plants are densely hairy throughout. The flowers are blue (rarely white) and the fruits have prickles about the edges. Stickseeds are usually found at the lower elevations in dry, disturbed habitats. *Lappula* is Latin, diminutive of lappa, meaning bur.

Field Notes/Uses No edible uses for the two of *Lappula* could be found. However, medicinally, a poultice from flatspine stickseed was applied to sores caused by biting insects, whereas as cold infusion was used as a lotion for sores and swellings. Additionally, the roots of European stickseed placed on a hot stone and allowed to smoke was used as an inhalant, and a snuff was made from the root for headaches

Gromwell (*Lithospermum/Buglossoides*)

General Description Two species of gromwell occur in Idaho, *L. ruderale* (western gromwell) and *L. arvense* (=*Buglossoides arvensis*) (corn gromwell). Western gromwell is a perennial herb up to 2 feet tall with a woody taproot and greenish yellow flowers. It is common in dry grassland areas and open forests up into the middle elevations. Corn gromwell is an annual plant about 2 feet tall with white or bluish white flowers. It is found in disturbed fields and pastures at low elevations. The genus name means "stone seed", referring to the hard nutlets.

Field Notes/Uses *Lithospermum* was used by Native Americans throughout the West as a medicine and food (Craighead et al. 1963). Although little is known about its chemistry, the effectiveness of several species as contraceptives and as depuratives for skin conditions warrants further investigation. Extracts of *L. ruderale* appear to contain a natural estrogen that interferes with hormonal balances in the female reproductive system. The Shoshone Indians used a tea made from the plant to treat diarrhea and as a female contraceptive, which caused permanent sterility after six months of continued use (Craighead et al. 1963). A salve from powdered and moistened leaves and stems was used for rheumatic and other pains where the skin is not broken.

Lungworts, Bluebells (*Mertensia*)

General Description There are eight species of *Mertensia* in Idaho. They are perennial herbs with succulent, alternate, and entire leaves. The blue flowers are funnel-form or trumpet-shaped. The genus is named after the German botanist, F.K. Mertens. The common name, lungwort, comes from a European species with spotted leaves, believed to be a remedy for lung disease.

Field Notes/Uses Bluebells are overlooked in many edible plant guides. The flowers can be nibbled upon raw or added to salads. Since the leaves are a bit hairy, we found them better when chopped up and added to soups. Weedon (1996) indicates that the rootstalk of *Mertensia ciliata* (mountain bluebell) is edible. However, bluebells may contain alkaloids and other constituents that can be toxic if consumed in large quantities (Tilford 1997).

Forget-me-not (*Myosotis*)

General Description Seven species of *Myosotis* occur in Idaho. In general, they are annual, biennial, or perennial herbs with alternate, entire leaves that are strongly veined. The flowers are blue, pink, or white, and are divided into a distinct tube and limb. The nutlets are smooth and shiny. The genus name is from the Greek, *mys*, meaning mouse, and *otis*, meaning ear, referring to the hairy leaves.

Field Notes/Uses The only reference to the uses of these plants was limited to *M. laxa* (bay forget-me-not). The plant was rubbed on the hair to act as a hair spray (Moerman 1998).

Popcorn Flower (*Plagiobothrys*)

General Description The three species in Idaho are annuals with alternate or opposite leaves. The white, salver-form flowers are in scorpioid racemes. The various species occur in moist soil.

Field Notes/Uses A rouge was obtained from the stem base and roots of some species of *Plagiobothrys* (Tull 1987). Many species have purple dye in the stems and roots. When pressed and dried in a folded sheet of clean paper, a mirror image pattern results (Spellenberg 1979, Schofield 1989). The shoots and flowers of related species in California (*P. fulvus*) were eaten, and leaves eaten as greens (Strike 1994).

Mustard Family - BRASSICACEAE

There are approximately 375 genera and 2,000 species in this family worldwide. Mustard flowers are easy to recognize with 4 petals in the form of a cross, hence the name "cruciform." There are usually six stamens, four of which are longer than the remaining two. The fruit is a pod, either long and thin (silique) or short and wide (silicle), with a partition down the middle dividing the seeds into two individual chambers.

Mustards are a friendly family, having a characteristic peppery taste. In general, the genera are safe to experiment with. Mustards are associated with the Roman practice of soaking seeds in newly fermented grape juice ("must"), drunk as a stimulant to prepare armies for battle. Cauliflower, turnip, radish, cabbage, rutabaga, and watercress are among the economically important plants in this family. Despite the high nutritional value of mustards and tenacity of plants, they are largely neglected. In America, mustards are often regarded as pests, but in Europe and Asia, various species are widely cultivated.

Madwort (*Alyssum*)

General Description These are small annual plants with foliage that appears dull gray due to the presence of dense, star-shaped hairs. The leaves are alternate and simple, and the flowers are short-stalked on the terminal portion of the stems. The flowers have light yellow petals that quickly fade to white. Fruits are egg-shaped or round in outline and have winged margins and a short style.

Field Notes/Uses Both species are Eurasian weeds and are widespread in distribution throughout most of the United States. The genus name is from the Greek *a* (without) and *lyssa* (rabies), to the supposed cure for rabies. The small leaves of pale madwort are mild tasting and can be eaten raw.

Rockcress (*Arabis*)

General Description These are biennial or perennial herbs with stellate hairs. Flowers are in racemes, usually white to purple in color. The fruits are linear siliques, usually flattened parallel to the partition. The many species of Arabis in Idaho are found in a variety of habitats at various elevations.

Field Notes/Uses The author has enjoyed forging on the first year's rosettes; a little bitter tasting but refreshing. The crushed plant of *A. puberula* (silver rockcress) serves as a liniment or mustard plaster (Train et al. 1957). Some species of rockcress were eaten by Native Americans, and an infusion was used to cure colds (Strike 1994).

Wintercress, Yellowrocket (*Barbarea*)

General Description The two species in Idaho are annual or biennial herbs with angled stems and pinnatifid leaves. The flowers are yellow, and the pods are linear and 4-angled. The genus is named for Saint Barbara. The Latin name is derived from the fact that the young leaves can be eaten on St. Barbara's Day in early December.

Field Notes/Uses The young stems and leaves of *B. orthceras* (American yellowrocket) and *B. vulgaris* (garden yellowrocket) can be eaten raw in salads or prepared as a potherb. We like to boil the plants in at least two changes of water.

Hoary False Madwort (*Berteroa incana*)

General Description This annual plant grows up to about 36 inches tall. The lower leaves are broadly linear and entire-margined with short petioles. These leaves eventually wither as the plant matures. The upper leaves are smaller and sessile. Foliage is grayish in color and a dense covering of branched hairs. Flowers occur at the terminal portion of the stem, and the notched petals are white. Fruits are slightly inflated and elliptical in shape, and have a prominent, persistent style at the tip.

Field Notes/Uses This is an introduced species from Europe usually seen in old field and along roadsides. The genus name, *Berteroa*, is for Carlo Guiseppe Bertero, a Piedmontese botanist. The species name, incana, means gray. The notched petals of the white flowers bear a resemblance to those of species of the genus *Draba*. However, this is a taller plant with leaves on the stem and an early spring flowering period. The round, ovate seed pods will distinguish this species from species of *Capsella*, *Lepidium*, and *Thlaspi* with flattened seed pods.

There is little information showing the importance of this plant species to wildlife. The plant produces many seeds, providing some feed value to songbirds and small mammals. Horses become intoxicated after eating green or dried plants. When mixed with alfalfa hay, *B. incana* can remain toxic for up to nine months. The toxic dose has not been determined.

Mustard (*Brassica*)

General Description Brassicas are large annuals with showy yellow flowers. The pods are round or 4-sided in cross section, with a conspicuous beak. In Idaho, four species can be found in waste places and fields at lower elevations. *Brassica* is the Latin name for cabbage.

Field Notes/Uses All species of *Brassica* have leaves that are edible as greens - an excellent source of Vitamin A, B, and C, calcium and potassium. The older leaves should be boiled in at least one change of water. The seeds contain thiocyanate and may cause goiter if consumed in large amounts (Weedon 1996, Harrington 1967). The seed can be ground or crushed to flour and applied as a mustard plaster. The plaster is a long-used remedy for aches and pains. Mustard oil is also a caustic irritant and can discolor and blister the skin if left on too long. The flower buds are rich in protein (Peterson 1977). The table mustard comes from *B. nigra* (black mustard). In China, mustard oil from *B. rapa* (field mustard) was used for illumination until the introduction of kerosene.

False Flax (*Camelina*)

General Description This is an introduced genus of annual or biennial herbs with all but one species, *C sativa*, occurring wild or as a weed. The leaves are entire or toothed, those on the stem sessile and clasping at the base. The yellow flowers are small and occur in rather long racemes.

Field Notes/Uses *C. sativa* has been cultivated since the Neolithic Age for the fibers of its stem and the edible oil contained in the seeds. It is said that an oil similar to linseed oil.

Shepherd's Purse (*Capsella bursa-pastoris*)

General Description This is an annual with leaves mostly in a basal rosette. The petals are white and the pods obcordate, strongly flattened contrary to the narrow septum. *Capsella* means "little box," referring to the fruit, as does the specific name *bursa-pastoris*. Collectively, the name means "purse of the shepherd."

Field Notes/Uses Shepherd's purse has been used as food for thousands of years. The seeds have been found in the stomach of Tollund man (approximately 500 BC - 400 AD) and during excavations of the Catal Huyuk site, approximately 5950 BC. The seeds of shepherd's purse may be parched and eaten or ground into flour. The whole pod, with the seeds beaten out, can be added to salads or soups or dried for winter use. The young leaves can be prepared as a potherb and are a good source of Vitamin C. With age, they develop a peppery taste. The entire herb (leaves, stems, green seed pods) can be chopped and added to soups. The roots may be ground or chopped and used as a ginger substitute. The seeds are known to cause blistering of the skin (Foster and Duke 1990).

The plant is extremely high in Vitamin K, the blood clotting vitamin. Mash or chew the leaves and hold them on a cut. The juice of the plant on a ball of cotton was used to stop a nosebleed. Shepherd's purse also contains significant amounts of calcium, potassium, sulfur, and ascorbic acid. Used as a decoction, shepherd's purse has been used to treat hemorrhoids, diarrhea, and bloody urine (Willard 1992). The decoction has a gentle detergent action, and is very cleansing to the skin.

Bittercress (*Cardamine*)

General Description The five species of *Cardamine* in Idaho are annuals or perennials with entire or pinnate leaves. The flowers are white or purple. The pods are elongate and flattened. The genus name comes from the Greek, because some bittercress were thought to have heart strengthening qualities.

Field Notes/Uses The plants can be eaten raw in salads, however, we suggest cooking them in at least a change of water to improve the taste. Some plants in this genus were reputed to have medicinal qualities that were used in the treatment of heart ailments.

Whitetop (*Cardaria*)

General Description These are rhizomatous perennials with erect or ascending stems and simple leaves. The herbage is pubescent with simple hairs, and the stalked flowers occur on the upper portion of the stems in a crowded flat-topped inflorescence. The fruits are egg-shaped to almost round and have a persistent style at the top.

Field Notes/Uses These species are aggressive perennials native to southwest Asia. They were likely introduced in multiple shipments of contaminated alfalfa seed from Turkestan into North America over a period of 40-50 years. All species readily establish in disturbed areas in range and wildlands and are favored during years of above average precipitation. Once established, colonies are difficult to eliminate because of deep, persistent roots. Cultivation can facilitate spread of plants by dispersing root fragments. However, repeated cultivation (bimonthly to monthly) can destroy colonies in 2-4 years. Flooding an area with 6-10 inches of water for two months can eliminate troublesome infestations.

The leaves of *C. draba* (whitetop) can be eaten raw or cooked. The young inflorescences resemble small broccoli and were eaten in much the same way.

Wild Cabbage (*Caulanthus*)

General Description: These are mostly annual or sometimes perennial herbs. Flowers occur in racemes and are white, purple, or yellow in color. There are approximately 18 species in the western North America. The genus name is from the Greek (*kaulos*), meaning stem, and (*anthos*), for flower - referring to cauliflower since some species can be used like it. Two species can be found in Idaho: *C. crassicaulis* (thickstem wild cabbage) and *C. pilosus* (hairy wild cabbage).

Field Notes/Uses The leaves of several western species were eaten by Native Americans.

Crossflower (*Chorispora tenella*)

General Description This is a leafy, taprooted winter annual. The plants branch primarily from the base, with some branching of the flowering stems, as well. The rosette leaves of seedling plants are deeply lobed or toothed and the mature plants are six to eighteen inches tall, with nearly unlobed leaves that are toothed or wavy. Flowering stems are leafy on the lower portions. Both leaves and stems are covered with gland-tipped hairs which make the plants feel sticky. The four-petalled flowers appear in early spring, and are purple to somewhat pinkish. Seed-pods are thin slightly curved, elongate and seedless (sterile) for about 1/3 of their length. Unlike the seed-pods of other mustards which split open along the length of the pod at maturity, the pods of this species break apart crosswise into two-seeded sections.

Field Notes/Uses This is a native of Russia or southwest Asia. It first was documented in this country in Lewiston, Idaho in 1929, and has spread throughout the western plains states, the western portion of the United States, and southern Canada. Crossflower likely was introduced into the United States by accident in imported seed, as is true with many members of the mustard family.

The genus name is from the Greek word *Chori*, meaning "separate" and *spora*, for "seed," and refers to the constricted seed shape. The species name of tenella means slender and probably refers to the overall plant or flower shape, both of which are slender.

Alkali Cusickiella (*Cusickiella douglasii*)

General Description This is a low, matted perennial with short stems 1 inch high, the stems covered with old leaves at the base. Individual leaves are thick, leathery, and rigid, where the earlier leaves are spatulate in shape and the later oblanceolate in shape. They are somewhat concave above with a stout midrib. The herbage consists of simple and forked hairs, or the leaves may occasionally be glabrous. The 2-10 flowers occur in racemes and are white with four petals. The fruits are somewhat inflated silicles which are ovoid and leathery. They are covered with short simple hairs. The species is also known as *Draba douglasii* (Douglas' draba).

Field Notes/Uses *Cusickiella* is a small genus containing two species of plants in the mustard family which are native to the western United States. Alkali cusickiella may be found on open rocky ridges among sagebrush and grasses. This is a Forest Service and Bureau of Land Management sensitive species.

Tansymustard (*Descurainia*)

General Description The four tansymustard species in Idaho are annual or biennial herbs with leaves that are 1-3 times pinnately divided. The foliage is covered with simple, branched, or short gland-tipped hairs. The flowers are cream-colored or light yellow and the pods are long, narrow, 3-sided to nearly round in cross section. These are weedy species occurring in disturbed soils at the lower elevations.

Field Notes/Uses All species of tansymustard are edible as greens, but are bitter. The seeds can be parched, ground and prepared as mush. The seeds were parched by tossing in a basket with hot stones or live coals, then ground into flour and made into mush. Because of its peppery taste, the mush was often mixed with the flour of other seeds to make it more palatable. Young leaves can be boiled or roasted between hot stones and eaten as green vegetables. The seeds were also used in poultices for wounds (Kirk 1975). However, one species, *D. pinnata* (western tansymustard) is reported to be poisonous in large quantities, causing blindness and paralysis of the tongue (Weedon 1996).

Draba (*Draba*)

General Description Drabas are annual, biennial, or perennial herbs with leafy or leafless stems and clusters of leaves at the base. The small flowers are white or yellow that fade to white with age. The pods are egg-shaped, elliptical, or club-shaped and sometimes twisted. Numerous species of *Draba* in Idaho can be found in open, dry to moist areas from the foothills to above timberline.

Field Notes/Uses The genus is also known as whitlow grass.The species are mainly unpalatable. Whitlow grasses were formerly used for treating "whitlows," inflammations of the fingertip, particularly next to the nail (Pojar and MacKinnon 1994).

Wallflower (*Erysimum*)

General Description Wallflowers are annual, biennial, or perennial herbs that are often tap-rooted. The herbage is covered with closely appressed forked hairs and yellow or orange flowers are showy. The linear pods are 4-sided in cross section with a small beak. The six species in Idaho can be found in dry to moist, open sites at mid to high elevations.

Field Notes/Uses Wallflowers were once used as a poultice. *Erysio* means to draw out, as in drawing out pain or causing blisters (Pojar and MacKinnon 1993). Additionally, an infusion of dried, pulverized *E. capitatum* (sanddune wallflower), was rubbed on the head and face, to prevent sunburn, or to alleviate heat exposure (Strike 1994). A pneumonia victim was cured by having his back massaged with chewed wallflower root.

Rod Halimolobos (*Halimolobos virgata*)

General Description This is a biennial species with one to several stems. The stems often branch and grow between 4-16 inches tall. The plant has a grayish appearance due to a covering of simple and branched hairs. The leaves taper to a slender basal stalk and may be pointed or blunt at the tip. The leaf edges may be smooth or wavy-toothed. The several leaves on the stem decrease in size going up the stem, but are widest at the base where they clasp the stem. The flowers open as they develop at the tip of the stem and occur on slender stalks. The sepals are purplish and the petals are white with pink or lavender veins. The fruits are cylindrical and upright at the ends of their stalks, and the numerous seeds are crowded in two rows on the pod which fits closely enough to make the seeds visible.

Field Notes/Uses Rod halimolobos is found only in western North America. In the United States the species is found in Montana, eastern Idaho, Wyoming, Colorado, Utah, Nevada and southeastern California.

This species can be distinguished from other members of the mustard family (Arabis) by its white flowers opening while still on the stem tip, by the presence of the many-branched hairs and by the fruit which is round in cross-section. Some *Arabis* may have white flowers but only flower buds are found at the stem tips. The flowers do not open until the stem has grown longer. Hairs are absent or unbranched on stems and leaves. Fruits of *Arabis* are usually somewhat flattened with smooth surfaces, which do not show the seeds inside.

Dame's Rocket (*Hesperis matronalis*)

General Description This is a tall herbaceous biennial or a short-lived perennial that is widely planted and escapes readily. The flowers are variable in color, ranging through many shades from white to pink to purple. It is frequently mistaken for "wild phlox," but the four petals and conspicuous fruits 2-5 inches long clearly identify it as a mustard family plant, rather than a phlox which would have 5 petals and much smaller inconspicuous capsules. It does best in moist or wet woods, but can tolerate a wide variety of other habitats.

Field Notes/Uses A native European, dame's rocket has naturalized through most of the United States. It can be long established in some of these "wild" situations, freely propagating by seed. The leaves, flower buds, and flowers are edible raw and great in salads. However, the plant does have diaphoretic and diuretic properties. The sale of dame's rocket to addresses in the state of Colorado is prohibited.

Peppergrass (*Lepidium*)

General Description The numerous species of peppergrass in Idaho are annual or biennial herbs with simple or 1-3 pinnately divided leaves that are alternate or basal. The flowers are white, yellow, or greenish, and the pods are flattened at right angles to the partition that separates the seed chambers. Peppergrass occurs in dry, open or vernally moist areas at low elevations.

Field Notes/Uses The young stems and leaves may be eaten raw or dried for future use. The plants contain Vitamins A and C, iron, and protein (Weedon 1996, Peterson 1977). The seed pods and seeds can be used as a flavoring. Fresh *L. virginicum* (Virginia pepperweed) plants were bruised or a tea made from leaves was used for poison ivy rash and scurvy (Foster and Duke 1990).

Note *Lepidium papilliferum* (sickspot peppergrass) is listed as endangered under the Federal Endangered Species Act.

Bladderpod (*Lesquerella*)

General Description These are perennial plants with stellate pubescence on the stems. They generally occur on dry slopes. The genus is named after Leo Lesquereux, a late 19th century American paleobotanist. There are about 40 species of annual to perennial densely hairy herbs that have small flowers. *Lesquerella* and *Physaria* are very similar and species of one genus could be mistaken for a member of the other. Currently, APG3 has determined that the genus *Lesquerella* is now defunct and species have been moved into the genus *Physaria*.

Field Notes/Uses Bladderpods have been developed into hundreds of ornamentals, and food plants such as cauliflower, cress, radish, kohlrabi, turnip, and rutabaga. The genus was named in honor of the early bryologist and paleobotanist Leo Lesquereux. Bladderpods are mostly a southwestern genus and there are about ninety species worldwide. The center of their range seems to be the southwestern United States and Mexico. To the casual observer, they may be mistaken for one of the more common mustards, but where most mustards have elongated pods, bladderpods are rounded.

The seeds of bladderpods contain oil that is similar to castor oil which could make it valuable as a commercial crop. Castor oil is used in many products such as fibers, paints, resins, lubricants, cosmetics, and hydraulic fluids. Since the 1950s the U.S. Department of Agriculture has been researching the domestication of *Lesquerella* and in the 1980s began breeding programs with various eastern U.S. species. A related species, Fendler's bladder pod, was used by the Navajo as a tea and used it as a remedy for the bites of spiders.

Watercress (*Nasturtium officinale*)

General Description Watercress is an aquatic herb with pinnately compound leaves, the 3-11 leaflets roundish to oblong, nearly entire and somewhat fleshy. The flowers are white and the pods usually about an inch long on rather long stalks. The plant is usually found growing in running water. It was formerly known as *Rorippa nasturtium-aquaticum*.

Field Notes/Uses The herbage of watercress is edible if the waters in which they grow are not polluted. However, finding unpolluted water may be difficult. One suggestion would be to soak the fresh greens in a disinfectant, or treat the water with water purification tablets, or a tablespoon of bleach in a quart of water. Then rinse the greens well in potable water to remove the chemicals. The peppery-tasting plants were eaten raw or cooked as a potherb. A good source of vitamins, watercress is listed as efficient in preventing scurvy. Medicinally, the plant was used for freckles, pimples, liver, and kidney troubles.

Bittercress (Cardamine brewerii) looks similar to watercress. To quickly differentiate the two species look at the fruits: bittercress fruits are linear and narrow; watercress fruits are round or 4-angled in cross section.

Daggerpod (*Phoenicaulis cheiranthoides*)

General Description This is perennial with a basal rosette of oblanceolate leaves with entire blades that narrow gradually to the long, slender petiole. The leaves are covered with numerous, fine to coarse hairs which range from cross-shaped to many-branched. The leaf margins are entire. The typically glabrous flowering stems have several sessile leaves with heart-shaped, clasping bases. The stems may be erect or prostrate. The stem leaves lanceolate and narrow in breadth. The inflorescence has many flowers in a raceme. The 4 sepals are pink or purplish in color and the pedicels are spreading. The pink to reddish-purple petals are obovate-oblanceolate in shape. The fruit is a broad, dagger-like silique; the siliques spread outwards at right angles to the stem. The surface of the silique is glabrous with a prominent ridge found the length of the middle of capsule.

Field Notes/Uses Daggerpod is found on rocky, thin soils from the sagebrush plains into ponderosa pine forests up to the cold, windswept ridgetops. It is an attractive wildflower that would probably be suitable for a dryland rock garden. The roots of *P. cheiranthoides* were used medicinally by Native Americans in California, although the method used is not reported (Strike 1994).

Twinpod (*Physaria*)

General Description Twinpods are tuft-forming perennials that have prostrate stems and clustered basal leaves with entire or toothed margins. The plants are covered with star-shaped (stellate) hairs. The flowers are yellow and occur on the terminal portion of the stem. The fruits are 2-lobed, inflated and have a notch at the top between the two lobes.

Field Notes/Uses In general, the twinpods are often difficult to distinguish from bladderpods (*Lesquerella*), but the inflated fruits of twinpods are divided by a notch into two balloon-like lobes. The fruits of bladderpods are nearly round with a slender style at the tip. The genus name is from the Greek physa (bladder), alluding to the inflated fruit.

Idaho twinpod (*P. didymocarpa* var. *lyrata*), also known as Salmon twin bladderpod, is endemic to low elevation foothills around the town of Salmon. It grows on steep south-facing slopes at elevations of 4,050 to 6,800 feet. Natural substrate is loose but fairly stable.

Wild Radish (*Raphanus sativus*)

General Description Wild radish is a branched herb up to three feet tall. The flowers are white with rose or purple veins, sometimes yellowish. The fruit is a rounded pod up to two inches long. Wild radish is a weed of waste places and fields at the lower elevations. The garden radish is a cultivated form of this species.

Field Notes/Uses The leaves, flowers, and pods of wild radish are used as a food, rather than the root. The root of wild radish is a bit too woody to be eaten like garden radishes. The flowers can be tossed in a salad or eaten alone as a snack. The fruits can be used in a salad, but must be collected before the seeds harden and the pods dry out. The taste of these pods resembles that of the garden radish.

Yellowcress (*Rorippa*)

General Description These are tap-rooted annuals or rhizomatous perennials with simple or pinnately divided leaves. The flowers are yellow or white; the pods are elliptical to linear and 3 sided to slightly compressed. The nine species in Idaho occur in moist, wet, or aquatic habitats up into the middle elevations.

Field Notes/Uses In general, *Rorippa* leaves can be eaten raw or can be cooked and flavored prior to eating. Medicinally, it was said that eating as much Rorippa as possible was good for treating liver and kidney problems. The plants are rich in iron, copper, calcium, sulfur, and magnesium. They also contain substantial amounts of vitamin A, B, B2, and C.

Mustard (*Sinapis*)

General Description Two species occur in Idaho: *S. alba* (white mustard) and *S. arvensis* (charlock mustard). The genus name derives from the Greek word synaptein meaning to fasten together. White mustard was formerely included in the genus *Brassica*, and can be used in much the same way as other *Brassica* species.

Field Notes/Uses The leaves of *S. arvensis* are edible at the juvenile stage of the plant. The seeds are considered to be poisonous and may cause gastrointestinal problems, especially if consumed in large quantities. The seeds of both species contain sinalbin (a thioglycoside) the cause of the the pungent taste. White mustard has fewer volitale oils than other species and hence milder.

Hedge Mustard (*Sisymbrium*)

General Description These are annual, biennial or perennial herbs. The small flowers are yellow or white, and the fruits are linear. The three species in Idaho are introduced from Europe and are widespread throughout the United States. They are usually found in waste places and disturbed habitats at low elevations.

Field Notes/Uses The seeds of *S. officinale* can be parched and then ground into a flour. The plants also make fine potherbs. As with other mustards, it is best to cook the plants in a couple of changes of water.

Prince's Plume (*Stanleya*)

General Description The four species in Idaho are annual or perennial herbs. The flowers are in elongated racemes. The fruits (siliques) are borne on a long stipe. The plants are usually found in sagebrush habitats at low elevations. The genus is named for Lord Edward Stanley, a British ornithologist who lived from 1775 to 1851.

Field Notes/Uses The tender leaves and stems of all four species can be prepared in much the same way as cabbage. They are bitter, but boiling in several changes of water remove some of the astringent properties. The seeds can be collected, parched, and then ground into a flour. They can be eaten as a mush or used in making breads.

Field Pennycress (*Thlaspi arvense*)

General Description Members of this genus are glabrous, annual or perennial herbs with leaves that have entire or toothed margins. The stem leaves have small lobes at the base that clasp the stem. The stalked flowers are borne on the bractless, terminal portion of the stems. Petals are white. The flattened fruits are egg-shaped to narrowly elliptical with a notch at the top and broad or narrow winged margins. This non-native species occurs in disturbed areas at the lower elevations, whereas the native species (now reclassified as *Noccaea*) occur in open areas up to the timberline.

Field Notes/Uses The young shoots of *T. arvense* can be eaten raw in salad, or cooked as a potherb. The plant has a mustard-like taste, with a hint of onion. Like other members of the mustard family, it's probably best to boil in at least two changes of water. Field pennycress is high in Vitamin C and contains a relatively large amount of sulphur (Harrington 1967). The species may be toxic in large quantities (Willard 1992).

Additionally, field pennycress is a popular food plant in various parts of the world, often being cultivated in Europe. It is used when the shoots are young and tender, utilized raw as a salad, or cooked as a potherb like spinach. To prepare, boil the shoots for 15 to 25 minutes and change the water once or twice. Even then a slight bitterness is present, so mix the greens with those from some blander tasting plants like pigweeds (*Amaranthus*). The young tender leaves, when used in a salad, are rather bitter tasting, so either mix with those from other plants or use a strong flavored salad dressing. Seeds and fruits have been used to flavor other food. But be cautious as field pennycress has caused illness when fed to cattle in hay.

Field pennycress was introduced to North America from Eurasia at a very early date, and is conceded to be a significant agricultural weed and competes keenly with crops for moisture and space, causing profound reductions in yield. Adding to the problem is seed distribution by spring floods. When once established, the soil soon becomes contaminated with its seeds. The plant bears an unpleasant odor, making it easy to identify. Because of its abominable smell it became widely known as stinkweed. Stinkweed is easily understood by anyone who has ever handled the weed, or tasted milk or butter from a cow which has eaten it.

Caution The entire plant of field pennycress is considered to be poisonous, but the seeds especially contain isoallyl thiocyanates and irritant oils. Symptoms include oral and gastrointestinal irritation, which leads to head shaking, salivating, colic, abdominal pain, vomiting and possibly diarrhea. Generally, symptoms do not occur unless large quantities are consumed over a period of time.

Sand Lacepod (*Thysanocarpus curvipes*)

General Description This is a slender branched annual with stem leaves and basal leaves arranged in a rosette. The flowers are purplish, and the circular, flattened pods are surrounded by a flat nearly circular wing. The species occurs in open areas at low to mid elevations.

Field Notes/Uses Sand lacepod seeds may be parched and eaten or ground into flour (Weedon 1996). A tea made from the plant is said to cure stomach-ache; a drink made from the leaf can be used to relieve colic (Strike 1994).

Watershield Family - CABOMBACEAE

Two genera and eight species in the Watershield family are distributed in temperate and tropical America, Africa, east Asia, and Australia. They are aquatic perennials typically found in freshwater. Only *Brasenia* is found in Idaho. The genus is sometimes placed in the Nymphaeaceae (Waterlily Family), but differs in having simple pistils.

Watershield (*Brasenia schreberi*)

General Description This plant is anchored to muddy substrates by slender rootstocks. All exposed portions of the plant are covered with a gelatinous sheath. The leaves are nearly round and arise near the tops of the stems. The flowers have purplish petals and sepals.

Field Notes/Uses The starchy rootstalk of watershield can be peeled, boiled, and eaten, or dried and stored or ground into flour. The unexpanded young leaves and leaf stems can also be eaten in a salad, slime and all (Coffey 1993, Weedon 1996). The rootstalks were used to cure dysentery and stomach-ache (Strike 1994).

Cactus Family - CACTACEAE

There are over 140 genera and 2,000 species within this family worldwide. Approximately 16 genera are native to the United States. Native to the Western Hemisphere, cacti have been spread all over the world, frequently carried by explorers and other travelers. Cacti are typically succulent spiny herbs of diverse form. One distinctive feature is the presence of areoles, which are round or elongated spots or openings that may be raised or pitted and usually arranged in rows or spirals over the surface of the plant. The spines grow from the areoles. The flowers have many sepals, petals, stamens, and an inferior ovary. Economic products from this family include ornamentals, edible fruits, nopalitas, and the hallucinogenic peyote. The spines of some cacti were once used as phonograph needles.

Foxtail Cactus (*Escobaria*)

General Description The cacti in this genus are ball-shaped (about $2\frac{1}{2}$ inches in diameter) and have greenish-white to deep red or purple flowers that are borne at the tip of the stem. The two species in Idaho include *E. missouriensis* (Missouri foxtail cactus) and *E. vivipara* (spinystar). They are usually found at the lower elevations in the foothills. Missouri foxtail cactus is often found along the Salmon River in Custer County. These species were once classified in the genus *Corypantha*.

> *Cacti are as evocative of the West as sagebrush. They have a number of distinctive shapes, their flowers are often large and attractive, and they have evolved over aeons of time to be perfectly at home in what we humans usually call "a hostile environment."*

Field Notes/Uses Like other species of cactus (i.e., *Opuntia*, etc.), the ripe fruits are edible. It is usually best to boil before eating. Spiny star is also found growing in Canada and more northern areas where other cacti do not. It has been suggested that these cacti are more "freeze tolerant" than its cousins. However, too much water appears to rot the plants, probably why it is found growing in the drier lower elevations.

Prickly Pear (*Opuntia*)

General Description Prickly pear need little introduction. In Idaho, there are four species which are succulent herbs with fibrous roots and the stems that are flat or cylindric. The leaves when present are small, fleshy, and awl-shaped. The genus name may be from the Papago name (*opun*) for this food plant, or named for a spiny plant of Opus, Greece. Many *Opuntia* species have glochids - minute, nearly invisible barbed hairs that grow in clusters in areoles. They easily become embedded in skin or clothing, and, because of their light tan or yellowish color and barbed surface, are almost impossible to remove. Prickly pear cacti occur in dry soils at the lower elevations.

Field Notes/Uses Cacti have provided Native Americans with food, medicine, dyes, and a variety of other uses for thousands of years. They have also been attributed with saving many lives by supplying both food and water to people stranded in the desert. Since eating too many cactus fruits can cause

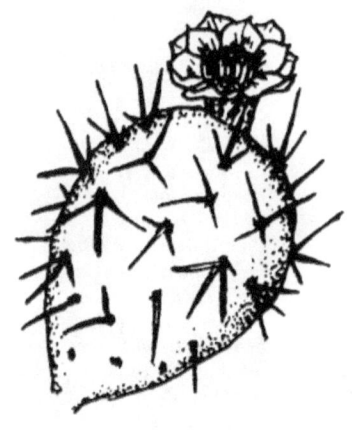

constipation, they should be eaten in moderation. The cactus also contains oxalic acid, so watch your intake as you may develop a deficiency of calcium.

The pads, especially younger pads, make an excellent cooked vegetable. Harvest the young pads by grasping them with tongs and slicing them at the stem joints. Hold over a flame to singe off the spines and glochids, and scrape off the remaining ones with a knife. Rinse well. Slice into thin strips and boil for at least 10 minutes. Drain off water, rinse to remove the slippery gum and they're ready to eat. To use the older pads, slice away the more fibrous section, and then cook accordingly.

After removing the spines, cactus fruits (aka prickly pears) can be peeled, and the pulp eaten raw, or boiled and then fried or stewed. One solution for removing the spines is to burn them off; another is to split the fruit into two halves and eat the insides. The pulp can also be sun or fire dried for future use. High in protein and oil the nutritious seeds may be eaten or dried and ground into flour. Grind the seeds into flour or add them to soups.

Prickly pear cacti pads have been used as a soothing poultice that can be applied to wounds and bruises (Willard 1992). A tea made from the flowers was said to increase urine flow, and a tea from stems was used as a wash to ease headaches, eye troubles, and insomnia. The fruits are high in calcium, potassium, and Vitamin C. Additionally, Native Americans in California used large baskets baited with crushed prickly pear pads to catch sardines. Another tribe roasted *Opuntia* stems, then soaked them. The resulting extract was used to improve the plasticity and cohesion of clay when making pottery (Strike 1994).

Archeologists in Texas have discovered purses made from prickly pear pads. According to Tull (1987), the dried pads were hollowed out to form a small container. A dye from the juice of the uncooked fruit can also be obtained. Bryan (1978) suggests letting wool soak for about a week in the fermenting juice. The color ranges from pink to magenta and appears to fade when exposed to sunlight.

Mountain Ball Cactus (*Pediocactus simpsonii*)

General Description This is a common and beautiful cactus, a globular type the reaches up to 6 inches in diameter and is strongly tubercled. It flowers from early May to June and the flowers are born in the center of the cactus and are usually a brilliant pink, but can sometimes be whitish in color. Interestingly, the flowers are generally pink on the eastern slope and yellowish on the western slope plants. The flowers usually remain closed on cloudy days. Another species, *P. nigrispinus* (snowball cactus), is also known to occur in Idaho.

Field Notes/Uses This genus contains eight species, seven of which are rare plants native to the Colorado Plateau region of Utah, Colorado, New Mexico, and Arizona. *Pediocactus* is derived from the Greek words *pedinos*, meaning plain/level and *kaktos*, meaning thistle. The species was first named by George Engelmann for army engineer James H. Simpson under the name *Echinocactus simpsonii*. Simpson led an expedition in Colorado, and Engelmann named the species "in honor of the gallant commander" of the expedition. The name of *Pediocactus simpsonii* was set forth by Britton and Rose, the genus name meaning from the plains -- even though this cactus grows in the mountains!

Water Starwort Family - CALLITRICHACEAE

Members of this family are small annual or perennial herbs with slender, usually lax stems. The leaves are simple, entire, and opposite or whorled. The minute unisexual flowers are borne in the axils of the leaves. There are no sepals or petals. The small, four-lobed fruit splits into four sections upon maturity. The plants are inconspicuous in standing water or drying mud. There is only one genus (*Callitriche*) and approximately 40 species in this family.

Water Starwort (*Callitriche*)

General Description Four species of water starwort are found in Idaho. *Callitriche* is the only genus of flowering plants in which aerial, floating, and subsurface pollination systems have all been reported. The genus name is from the Greek, *kallos*, which means beautiful, and *trichos*, which means hair, referring to the slender stems.

Field Notes/Uses Strike (1994) indicates that Native Americans in California used *Callitriche* to relieve urinary problems. The method is not reported, however.

Bluebell Family - CAMPANULACEAE

Worldwide, there are over 70 genera and 2,000 species in this family. Twelve genera are native to the United States. These are annual or perennial herbs usually with milky juice. The flowers are typically 5-parted with the calyx divided into separate sepals, and the corolla is five lobed and bell-shaped. The family is of little economic importance, but some species are cultivated as ornamentals.

Harebell, Bluebell (*Campanula*)

General Description These are perennial herbs from a rhizome. The blue (occasionally white) flowers are tubular-, bell- or cup-shaped. The genus name is from the Latin "bell," and the common name, harebell, may allude to an association with witches, who were believed to transform themselves into hares, porters of bad luck when they crossed a person's path. The four species in Idaho can be found in open, dry, or rocky areas from low elevations to above timberline.
    Field Notes/Uses The leaves and shoots of at least *C. rotundifolia* (bluebell bellflower) can be used in salads or cooked as a potherb. The roots can also be boiled and eaten, and have a nut-like taste. *C. rapunculoides* (Rampion Bellflower) is also edible (Harrington 1967).

Clasping Venus' Looking-glass (*Triodanis perfoliata*)

General Description This is an upright annual reaching a height of 2½ feet. It grows in disturbed soil or moist areas such as ditches and along roadways. The blue flowers vary in width, ranging from 1/4 to3/4 inch in diameter. An older scientific name for this plant is *Specularia perfoliata*.
    Field Notes/Uses When it is sunny, the flowers open up during the morning, and remain open for the rest of the day. They are attractive, but rather small. The common name of this plant probably refers to the shiny seeds of a related European species. The seeds of Venus' looking glass are too tiny to appear shiny to the unaided human eye.
    The species was used by the Cherokee as a drug plant in a treatment for indigestion. Additionally, an infusion of the roots was taken and used as a bath for dyspepsia. Venus' looking glass is sometimes used in dried flower arrangements.

Hemp Family - CANNABACEAE

The Hemp family consists of annual herbs or climbing perennial herbs. The leaves are opposite, simple or compound. There are two genera (i.e., *Cannabis* and *Humulus*), with 3-5 species distributed in the north temperate zone, and they are widely cultivated. *Humulus* is native to the United States. The family is a source of hempen fiber, oils, edible seeds, hops, and tetrahydrocannabinols (THC), the psychoactive compound in *Cannabis*. Some references place the family in the Moraceae (Mulberry Family).

Hemp, Marijuana (*Cannabis sativa*)

General Description This is an unbranched, coarse, aromatic annual. The lower stem leaves are opposite and the upper ones are alternate. The leaves are palmately compound with 3-9 leaflets that are lance-shaped to elliptic, coarsely toothed. The small, green male and female flowers (staminate and pistillate flowers, respectively) are found on separate plants. The fruit is an achene which is enclosed in the calyx and covered by a persistent bract.

Field Notes/Uses The plant is a native of Asia and was cultivated in Europe for fiber (hemp) and seeds (hemp butter, oil). The seeds can be used as food, parched and mixed into a batter and fried into cakes (Harrington 1967). Kephart (1964) indicates that the young shoots were used as a substitute for asparagus in Belgium, but because the plant contains poisonous alkaloids it should be avoided.

While the plant is now illegally cultivated in Idaho and elsewhere for its euphoria-inducing properties, marijuana has been legitimately used to treat glaucoma, and relieve nausea following chemotherapy (Foster and Duke 1990). Although much maligned, marijuana is potentially a very useful medicinal plant.

Warning Cultivation of this species and use as a drug is forbidden by law.

Common Hop (*Humulus lupulus*)

General Description The genus was formerly included in the Mulberry family (Moraceae). Common hop is a strongly twining, herbaceous vine with stems up to 15 feet long. The stems and leaves are rough to the touch. The leaves are opposite, serrate, 3-7 lobed, with heart-shaped bases. The underside of the leaves is glandular. The flowers are small, green, and unisexual. This is a widely cultivated plant from Europe and Asia.

Field Notes/Uses Hops are primarily grown for their fruits, used in brewing to give ale and beer a distinctive bitter taste. Additionally, the young shoots can be prepared as potherbs. They can be boiled in water for about 3-5 minutes, and then boiled again in fresh water until tender.

A tea from the fruits was traditionally used as a sedative, antispasmodic, and diuretic, and used for insomnia, cramps, coughs, and fevers. Externally, the tea was used for bruises, boils, inflammations, and rheumatism (Foster and Duke 1990). Recently, clinical studies have disproved the sedative qualities of hops and found that it has no physiological activity on the nervous system; yet, anyone who drinks much of the tea tends to fall asleep or become groggy (Moore 1979).

Caution The plant is known to cause dermatitis when handled.

Caper Family - CAPPARACEAE

Members of this family are shrubs, trees, or rarely herbs. They have simple or palmately compound leaves. Flowers have 4 sepals and 4 petals with the fruit a capsule or berry. There are about 46 genera and 800 species distributed in the tropical and subtropical areas of the world. Eight or nine genera are native to the United States. The family is of economic importance as a source of ornamentals and capers, a salad seasoning.

Bee Plant (*Cleome serrulata*)

General Description This is one of three species that occur in Idaho. It is an erect, showy plant up to 40 inches tall with alternate leaves divided into three lance-shaped, entire leaflets. The reddish-purple to pink flowers are arranged in a dense, narrow, terminal inflorescence. The petals are separate while the sepals are united. The fruits are long-stalked, pendulous capsules, linear to lance-shaped in outline. Bee plant is found in disturbed areas (i.e., roadsides, railroad right-of-way) at low elevations.

Field Notes/Uses An important food for many western Native Americans, bee plant was extensively used as a potherb. The young tender shoots and leaves, and flowers are preferred. The plant has an unpleasant odor, especially when older, and a pungent taste much like the mustards. We found it necessary to cook the plants in at least two changes of water to remove the bitter taste. The seeds can also be collected and ground into flour.

The Blackfeet Indians used the whole plant to make a medicinal tea to alleviate fever (Johnston 1970). Bee plant may have been used by Native Americans to treat stomach-aches (Strike 1994).

As a dye, plants are collected in quantity and boiled down for several hours until a thick, fluid residue is produced. The water is then drained off and the plants allowed to dry and harden into cakes. When black dye or paint is needed, a piece of the cake is soaked in hot water.

Clammyweed (*Polanisia dodecandra*)

General Description This is a sticky, hairy annual with simple stems and a strong, rank odor. The leaves are 2 inches long and bear 3 leaflets about an inch long. There are about 20 flowers clustered at top of plant, each flowers about a ½ inch long, white with purple bases; there are numerous stamens that overtop the petals. The fruits consist of slender capsules about one or two inches long that are filled with many tiny, dark seeds.

Field Notes/Uses The name clammyweed is in reference to the glandular, sticky pubescence that covers the plant. Touching the foliage gives the fingers a clammy feeling. Bees visit the flowers for nectar and while there are flies that feed on the pollen, however, they do little in effectively pollinating the flowers. The plants are not known to be toxic even though they emit a foul odor.

Honeysuckle Family - CAPRIFOLIACEAE

There are 15 genera and 400 species in the Honeysuckle family. Of the 15 genera, 7 are native to the United States. They are woody plants with opposite leaves. The flowers are 5-merous, with the petals fused, and an inferior ovary. Many genera in this family are cultivated as ornamentals. In recent years, members of this family have been moved to other families.

Twinflower (*Linnaea borealis*)

General Description Twinflower is a slender, trailing, mat-forming evergreen with short, leafless branches that divide into two at the top. At the top of these small branches arise small bell-shaped pink or white flowers, hence the name twinflower. The evergreen leaves are oval or round and about ½ inch long. Twinflower is found in the middle to subalpine elevations and is associated with conifers and moss-covered sites that also support *Pyrola*, *Clintonia uniflora*, and *Chimaphila umbellata*.

Field Notes/Uses The genus name honors Carolus Linnaeus of Sweden, the person largely responsible for developing the binomial (two name) system of naming plants and animals. It is said that twinflower was Linnaeus' favorite flower. The species name, borealis, means northern.

Twinflower is known to some Native

Americans for its medicinal uses. As an orthopedic aid, the plant was mashed for inflammation of the limbs, whereas a poultice of the whole plant was applied to the head for headaches. The Algonquin Indians of Quebec made an infusion of the entire plant for menstrual difficulties and as a means to ensure "good health of the child" for pregnant women. The Iroquois made a decoction of the plant and use it as a sedative for crying children, and for children with cramps or fever (Moerman 1998).

Honeysuckle (*Lonicera*)

General Description The six *Lonicera* species in Idaho are shrubs or woody vines with entire and opposite leaves. They can be found in a variety of habitats from the foothills up to the alpine zone. The genus is named for Adam Lonitzer, a German naturalist who lived from 1528-1586.

Field Notes/Uses The berries of honeysuckle are seedy, but can be eaten raw, or dried for future use. However, Pojar and MacKinnon (1994) believe the orange berries of *L. ciliosa* (orange honeysuckle) were not eaten and may be poisonous, and *L. involucra* (black twinberry) berries were not considered edible by most people in the Northwest.

The bark and twigs of *L. involucra* were used for a variety of medicinal preparations, ranging from digestive tract problems to contraceptives. Additionally, the juice from the stems was used as an antidote for bee stings.
The berries provide a black pigment. The long stems of honeysuckle were used as basket foundation material by a number of Native American tribes. They also peeled and split the hairy stems as wrapping material for coiled baskets.

Elderberry (*Sambucus*)

General Description Elderberries are shrubs with pithy stems. The two species in Idaho have large, compound leaves with serrated leaflets. The white flowers are arranged in dense clusters. The fruits may be red or blue-black. Elderberries can be found in open areas, hillsides, and riparian habitats in the montane zone. The genus name comes from the Greek *sambuke*, an instrument made from the hollow stem.

Field Notes/Uses The blue or black elderberry berries of *S. nigra* (black elderberry) are edible raw, or they can be made into excellent jams, jellies, and wine. They can also be dried and stored for winter use. The seeds contain hydrocyanic acid, and if eaten in quantity can cause diarrhea and nausea. It is best to cook the berries or strain the seeds before use. The red-berried species contain much higher concentrations of these compounds and should be considered poisonous.

The blossoms can be added to pancakes to lighten batter and add flavor. The dried flowers were also ground and added to flours and baking mixes. Flower buds can be pickled or steamed as a potherb. Both the flowers and fruits contain a rich source of Vitamin C (Strike 1994).

The fresh flowers can be used externally as a decoction for an antiseptic wash (Willard 1992). Flower tea contains a natural estrogen and is often effective for relieving menstrual cramps (Strike 1994). The leaves were used as poultices for sprains and skin irritations. The leaves and flowers were common ingredients in skin salves for piles, burns, and boils. Recent studies of elderberry have confirmed that the berries possess antiviral properties that may be useful against influenza (Tilford 1997).

Elderberry stems can be cut and dried for use as musical instruments. After drying, holes can be bored into the branches to make flutes. During the drying process, the poisons are said to dissipate. The stems can also be used in making bows and arrow shafts in hunting small game. The odorous leaves can be used in water and sprayed on plants to repel aphids. The pith of the stem is used by watchmakers to absorb grease and oil. The leaves, with chrome as a mordant, yield a green hue. The berries, with alum and cream of tarter, will yield a crimson dye.

CAUTION The seeds, leaves, bark, and roots contain hydrocyanic acid and an alkaloid sambucine. They are toxic and cause acute emetic and laxative effects (Tilford 1993). Berries should be consumed when ripe and used for food after cooking and removal of seeds.

Snowberry (*Symphoricarpus*)

General Description Snowberries are erect shrubs with elliptical to egg-shaped leaves. Flowers are white to pink and bell-shaped, accompanied by two small bracts. The fruits are berry-like and white. Six species of snowberry in Idaho are found in dry soils at various elevations.

Field Notes/Uses The white, tasteless berries are edible raw or cooked, and are said to be emetic and cathartic in large amounts (Kirk 1975). Saponin is found in the leaves and can be used as a natural cleaning agent. A decoction of the pounded roots has been used for colds and stomachache (Sweet 1976). An infusion made from *S. albus* (common snowberry) was used to cure sores and skin lesions, and a root decoction was used to alleviate colds and stomach ailments (Strike 1994).

High-bush-cranberry (*Viburnum*)

General Description The two species in Idaho are shrubs with opposite leaves that are lobed or toothed. The flowers are white, showy, and borne in a large umbrella-like inflorescence. They can be found in moist areas and along streams.

Field Notes/Uses The berries of *V. edule* (squashberry) and *V. opulus* (American cranberry bush) are edible and taste somewhat like cranberries.

Pink Family - CARYOPHYLLACEAE

There are approximately 80 genera and 2,000 species in this family found in the north temperate zones. About 20 genera are native to the United States. In general, they are annual or perennial herbs with opposite, simple leaves. The stems are often swollen at the joints. The flowers are 5-merous, and the calyx is tubular or has distinct sepals. Petals are often deeply notched, appearing like ten petals. Some species are cultivated as a source of ornamentals, and several genera are regarded as weedy.

"Sandworts" Complex (*Arenaria, Eremogne, Minurtia, Moehringia*)

General Description The genus *Arenaria* underwent some changes and some of the species were placed in new genera. The species in this genus are generally annual or perennial herbs with opposite leaves. The white flowers are borne in open to congested, flat-topped inflorescences. The six species in Idaho are found in various habitats from the foothills to above timberline in dry, rocky, and open areas, as well as moist open forests. The genus name is from the Latin, arena, referring to sand, the habitat of many species.

Field Notes/Uses The sandworts are known for their medicinal uses by several Native American tribes. The roots of *A. aculeata* (prickly sandwort) were used as a decoction by the Shoshone as eyewash, whereas a poultice of steeped leaves of *A. congesta* (ballhead sandwort) was applied to swellings.

Chickweed (*Cerastium*)

General Description In Idaho the six species of chickweed are annual or perennial herbs with opposite leaves. The herbage is usually hairy and sticky. The flowers are white and the petals are deeply lobed at the tip. The fruit is a cylindrical capsule, often slightly curved at maturity. The annual species are found at lower elevations; the perennial species occur in dry to moist, open habitats up to the alpine zone. The genus name is from the Greek (*keras*), meaning horn, referring to the tapered capsule, which in some species is bent slightly like a cow's horn.

Field Notes/Uses *Cerastium* is frequently confused with *Stellaria media* (chickweed), but to the general forager there is no danger. The tender leaves and stems of most *Cerastium* can be added to a salad, but we found they are better if boiled first and served as greens.

Jagged Chickweed (*Holosteum umbellatum*)

General Description This annual has simple or basally branched stems typically less than 8 inches tall and usually much smaller. The basal leaves are oblong lance-shaped and ¼ to ¾ inch long. The 2-3 pairs of stem leaves are wider. The foliage is soft hairy below and glandular hairy above. Flowers are borne in a flat-topped inflorescence at the top of the stem. Sepals are separate and about 1/8 inch long. The white petals are jagged at the tips and slightly longer than the sepals. The seed capsule is nearly twice as long as the sepals. Jagged chickweed is common in moist or vernally moist, open habitats in the valley and montane zones.

Field Notes/Uses In years with a wet spring, the plant appears in great masses in overgrazed pastures. It is introduced from Eurasia and is well established throughout much of North America.

Tuber Starwort (*Pseudostellaria jamesiana*)

General Description This is a weak-stemmed, glandular perennial with lanceolate leaves and many few-flowered cymes of white flowers that have slightly 2-lobed petals. This species is found about meadows and damp places and flowers May to July. *Pseudostellaria* refers to starwort's resemblance to the genus *Stellaria*.

Field Notes/Uses The species name, *Jamesiana*, is for Edwin P. James (1797-1861), an American naturalist and botanical explorer in the Rocky Mountains. He studied medicine, then learned botany from Professor John Torrey, and in 1820 became the naturalist-surgeon of the federal government's Yellowstone Expedition which explored the Rockies all the way south into New Mexico. He and two colleagues were the first Americans to ascend Pike's Peak, and he was the first plant collector to explore the high alpine regions of the Rocky Mountains.

The tuber-like swellings of this species can be eaten raw or dried in the sun. They have a thin, light brown rind, and a tender rather mealy texture inside, similar to a potato.

Bouncing-bet, Soapwort (*Saponaria*)

General Description This an erect perennial herb with sessile or nearly sessile leaves. The flowers are showy, usually pale pink. Soapwort can be found along roadsides, disturbed areas, and waste places at the lower elevations. The plant has escaped from cultivation. The genus name is Latin for soap, since the juice of the plant lathers with water.

Field Notes/Uses The plant contains saponins and will irritate the digestive tract if eaten. The crushed green plant and roots can be used as a soap substitute.

Catchfly (*Silene*)

General Description Many species of *Silene* occur in Idaho. They are annual, biennial, or perennial herbs with opposite leaves. The sepals are united and often inflated into a 5-lobed tube. The petals are lobed at the tip and have appendages at point where the broader upper portion (blade) joins the narrower lower segment (claw). The various species are found in a variety of habitats.

Field Notes/Uses The young shoots of S. acaulis (moss campion) and S. cucubalus (=S. vulgaris) (bladder campion) can be used as potherbs. Moss campion can be found in rocky, open, exposed habitats above timberline. Bladder campion grows along roadsides and gravelly riverbanks at low elevations. The older aboveground plant of moss campion is also edible, but should be boiled until tender (Peterson 1977, Kirk 1975).

The sap of S. antirrhina (sleepy silene) was used by Native Americans in California to paint designs on the faces of young girls. The designs were cosmetic, not ritualistic (Strike 1994).

Sandspurry (*Spergularia*)

General Description *Spergularia* can be distinguished from other similar small Caryophyllacious plants by presence of stipules - small appendages at bases of leaves. Three species occur in Idaho.

Field Notes/Uses The tiny seeds are sometimes eaten, but contain saponin. They were gathered, ground up, and mixed with flour to make bread.

Chickweed, Starwort (*Stellaria*)

General Description The many species of *Stellaria* in Idaho are described as low, annual or perennial herbs. The five petals are separate to the base, white, and deeply lobed. They occur in various habitats up to timberline. One species, S. media, is the common chickweed found in most gardens and lawns.

Field Notes/Uses The young shoots of common chickweed can be used as salad herbs or potherbs if cooked like spinach. Although it is edible raw, we prefer to boil for a few minutes before eating. Since the plants are usually quite small and only the youngest parts are good, chickweed can be tedious to collect.

The greens are low in calories and packed with copper, iron, phosphorus, calcium, potassium, and Vitamin C - valued in the prevention and treatment of scurvy (Tull 1987). The edibility of the other species of *Stellaria* is unknown.

Medicinally, *S. media* can be used as a tonic, in large quantities a laxative, and diuretic. For itchy skin, make a strong tea and wash the area (Foster and Duke 1990). A poultice of the plant has been used to treat skin sores, ulcers, and infections as well as eye infections and hemorrhoids (Tilford 1993, Willard 1992).

Cow Soapwort (*Vaccaria hispanica*)

General Description This is a garden escapee and crop weed, recorded from wasteland and crops in the southwest. It is an erect annual, up to 24 inches tall. It has pink petals and hairless, blue-green leaves. This is a native plant of Europe.

Field Notes/Uses This herb contains a variety of saponins, mainly including vascegoside, through the hydrolysis of which vaccaroside is obtained, through the further hydrolysis of which yuccagenin is obtained. It also contains vaccaric flavonoid glycoside, alkaloids and such compounds as coumarin. The seed is anodyne (relieves pain), discutient (disperses morbid matter), diuretic, emmenagogue (promotes menstrual discharge), galactogogue, styptic (contracts or binds) and vulnerary (heals wounds). A decoction is used to treat skin problems, breast tumors, menstrual problems, deficiency of lactation and sluggish labor. The seeds are also taken internally as a galactogogue.

Staff-tree Family - CELASTRACEAE

These are shrubs with small inconspicuous flowers borne in the axils of the leaves. The 4-5 sepals are united at the base, and the petals are separate. The fruit is a capsule. There are 60 genera and 850 species distributed in tropical and temperate regions of the world. Ten genera are native to the United States. Several species are cultivated as ornamentals.

Oregon Boxleaf (*Paxistima myrsinites*)

General Description This is a low, dense, evergreen shrub. The thick, leathery leaves are opposite, oval to elliptic in shape, and the toothed margins are slightly rolled under. The small flowers are maroon in color. The plant is found in coniferous forest, rocky openings, and dry mountain slopes from low to mid elevations. The genus is often spelled *Pachistima*.

Field Notes/Uses The fruits were eaten by some Native Americans in California (Strike 1994), while other sources indicate that the plant is inedible. Additionally, they used the boiled leaves to poultice pain or inflammation (Strike 1994).

Hornwort Family - CERATOPHYLLACEAE

The Hornwort family has only one genus, *Ceratophyllum*, with approximately three species distributed worldwide.

Coon's Tail (*Ceratophyllum demersum*)

General Description This is a rootless, submersed or free-floating aquatic forb with slender, lax, and much branched stems. The sessile leaves are in whorls of 5-12, and the blades are dissected into linear, filamentous segments whose shape varies with the position on the plant. The minute flowers have no petals, and are borne in the axils of the leaves. Coon's tail is a common plant in standing or slowly flowing water of rivers, sloughs, and ponds at low to mid elevations.

Field Notes/Uses Native Americans in California used coon's tail to make a soothing lotion that was used on sore or inflamed skin (Strike 1994).

Goosefoot Family - CHENOPODIACEAE

Approximately 102 genera and 1,500 species in the Goosefoot family are found worldwide. Fourteen genera are native to the United States, mostly in the West. This family includes several food plants (e.g., beets and spinach) and weeds (e.g., Russian thistle). Recent studies have merged members of this family with the Amaranth family (Amaranthaceae).

Iodinebush (*Allenrolfea occidentalis*)

General Description The bumpy green stems are fleshy and appear jointed at the internodes between segments. Often the segments are so short they are nearly round. The leaves appear as flaky scales scattered across the surface of the stems. It grows in sandy, often salty, distinctly alkaline soils, such as desert washes and saline dry lakebeds.

Field Notes/Uses This plant has many common names such as iodine bush, bush pickleweed, or kern greasewood. The name iodinebush was probably given to this plant because when you crush its leaves, a brown color emerges. This genus is named for the botanist Allen Rolfe from Kew Gardens in London around early 1900's. The seeds of this plant can be parched, ground into flour, and mixed with water to make a drink or mush. To some Native Americans this was considered a starvation food

Saltbush (*Atriplex*)

General Description Many species of *Atriplex* occur in Idaho. They are annual or perennial herbs or shrubs with alternate leaves, and glabrous or scaly herbage. The flowers are unisexual, and individual plants have one or both sexes. The various species are found at the lower elevations in valleys, disturbed areas, or in dry, alkaline soils.

Field Notes/Uses The many uses for these plants include food to medicine and dyes, as well as soap and spice. The young leaves of many species can be cooked and eaten as greens and have a very distinct salty taste. We've often added them to otherwise bland foods to make our wild meals less boring. Add the leaves to meats while cooking will help spice them up. The seeds were parched, ground into flour, and made into mush. They can also be soaked in water for a few minutes to make a rather pleasant tasting drink. The Navajo used the flowers to make puddings. The ashes of *A. canescens* (fourwing saltbush) make a good substitute for baking soda.

Medicinally, the various species provided the Native Americans with many uses. As an analgesic, the Navajo used the leaves of *A. argentea* (silverscale saltbush) as a fumigant for pain, and the Zuni made a poultice from the chewed root for application to sores and wounds. A warm poultice made from the pulverized root fourwing saltbush was used to treat toothaches.

The leaves and roots of many species were used as soap. They were rubbed in water for lather and used in washing clothing and baskets. Many Native Americans also carved arrowheads from the wood for use as weapons and for hunting. The seeds of some species were also used in making a black dye.

Common Kochia (*Bassia scoparia*)

General Description Formerly in the genus *Kochia*, there are five species including common kochia, the more likely species to be encountered and foraged for. Common kochia is a bushy annual with stems up to three feet tall. The leaves are alternate, narrowly lance-shaped and tapered at both ends. The herbage may or may not be covered with hairs. Flowers are solitary or in clusters in spikes. The species is common in open, disturbed habitats at low elevations.

Field Notes/Uses Common kochia is native to Europe and was introduced into the United States as an ornamental. It has since escaped and has become well established. In Asia, Japan, and China, common kochia was cultivated for its seeds. The tips of the young shoots can be prepared as potherbs. The seeds can be eaten raw or cooked, or ground into meal and used in bread making.

Goosefoot (*Chenopodium*)

General Description The many species of *Chenopodium* in Idaho are annuals with mealy foliage. The small green flowers are borne in dense clusters in the leaf axils. Most species occur in disturbed areas at the lower elevations or wetland and riparian habitats up to the alpine zone. The genus name comes from *Cheno*, meaning goose, and *podium*, meaning foot, because the triangular leaves resemble the shape of a goose's foot. Oil of chenopodium, distilled from the fruits, contains a broad-spectrum vermifuge that is widely used in veterinary medicine.

Field Notes/Uses The leaves, tops, and seeds of all species can be used as an emergency or basic food and are quite tasty and nutritious. High in protein, the greens are a good source of Vitamins A and C, iron, potassium, and are extremely rich in calcium (Tull 1987). Since it does not become bitter with age, both young and old plants can be used. Leaves may be used raw in salads or boiled in water like spinach. The water can be saved and used as a yellow dye. The leaves were also eaten to treat stomachaches and prevent scurvy. A leaf poultice was used on burns (Foster and Duke 1990). The flower buds and flowers can be used as potherbs. A single plant can produce up to 70,000 seeds. Seeds can be ground as flour for use in bread or cooked as mush. Seeds can also be eaten without grinding or incorporated into pinole (flour made from a mixture of seeds of small plants). The seeds contain about 15% protein and 55% carbohydrates, more than is found in corn (Schery 1972). The seeds can also used as a coffee substitute.

Large quantities of the plant should not be eaten as many species contain high levels oxalic acid which tends to bind calcium and prevent its proper absorption into the body (Tull 1987). Additionally, *Chenopodium* has been known to accumulate toxic levels of nitrates and may cause livestock poisoning. But, because large quantities of the plant must be consumed to cause problems, this type of poisoning may be unlikely. Cooking or freezing of Chenopodium apparently breaks down the oxalic acid.

The hard root of some *Chenopodium* species was stored until needed, then grated on a rock to make soap. The leaves were also used to make soap, but are not as effective as the roots (Strike 1994).

Winged Pigweed (*Cycloloma atriplicifolium*)

General Description What is most distinctive of this plant in its sessile, flat, saucer-like or wheel-like winged fruits and its sinuate almost holly-like leaves with sharp-tipped lobes. The young stems and leaves are loosely tomentose. In the fall the plant, especially the fruit, may become a rich purple-red, as in some other members of the family. The whole mature much-branched, dome-shaped plant is fine tumbleweed.

Field Notes/Uses This is another one of those "tumbleweed" species, where the small taproot easily breaks free from the fine fibrous roots, sending the round, spidery top across an open landscape, scattering seed along the way. The seeds were used to make flour and then mush.

Jerusalem Oak Goosefoot (*Dysphania botrys*)

General Description Formerly known as *Chenopodium botrys* and as *Ambrosia mexicana*. The genus *Dysphania* is known as the glandular goosefoots. Jerusalem oak goosefoot is native to the Mediterranean region.

Field Notes/Uses The leaves and seeds are more or less edible. Leaves can be cooked or raw leaves should only be eaten in small quantities. The seeds can be ground into a meal and used with flour in making bread. The seed is small and fiddly, it should be soaked in water overnight and thoroughly rinsed before it is used in order to remove any saponins. The leaves are a tea substitute.

A gold or green dye can be obtained from the whole plant. Additionally, the dried plant is a moth repellent. The whole plant is very aromatic and is used as a scent in pillows, bags, and baskets.

Spiny Hopsage (*Grayia spinosa*)

General Description This is a low mealy appearing shrub with stiff, spreading spine-tipped branches and slightly fleshy leaves with gray tips becoming pinkish with age. The flowers occur in heads, these borne in terminal or axillary spikes or panicles. The fruits are closely subtended by a pair of attractive rose-purple, thin, flat-winged bracts that are united to the middle or higher.

Field Notes/Uses The genus *Grayia* is named for the American botanist Asa Gray, and contains a single species. This plant is extremely drought tolerant thanks to a deep root system. It serves as a good food source for browsers, especially in the dry months when other plants have dropped their leaves. Some Native American people traditionally ground parched seeds of spiny hopsage to make pinole flour.

Winterfat (*Krascheninnikovia lanata*)

General Description Winterfat is a small shrub found at the lower plains and foothills elevations, often in saline or alkaline areas. The leaves are alternate, narrow and entire, whereas the flowers occur in heads or spikes in the axils of the leaves.

Field Notes/Uses While the edibility of this species is unknown, it is considered an important forage plant for horses and other livestock. Medicinally, the plant has been used by many Native American tribes. For example, the Hopi Indians used the powdered root for burns, and a decoction of the leaves was used for fevers. The Navajo made a poultice of the chewed leaves and applied it to a poison ivy rash. The Navajo also incorporated the stems and leaves of this plant in sweat house ceremonies by placing them on hot rocks for the Mountain Chant (Moerman 1998).

Povertyweed (*Monolepis nuttalliana*)

General Description While there are two other species in the state (*M. pusilla* and *M. spathulata*), povertyweed is more wide ranging and likely to be encountered. It is a low growing winter annual with prostrate or ascending stems. The leaves are somewhat succulent and lance-shaped, broadened and lobed at the base. Flowers are borne in dense clusters at the leaf bases and the solitary sepal is reddish in color. The seeds are dark brown. The plant is found in open disturbed habitats at the lower elevations.

Field Notes/Uses The above ground parts of povertyweed may be eaten as a potherb. The seeds are also edible. It is unknown if the other two species can be used similarly.

Red Swampfire, Pickleweed (*Salicornia rubra*)

General Description These are fleshy, hairless, herbaceous annual plants that have leafless, jointed stems bearing opposite branches. The flowers are borne in fleshy, cylindrical spikes with the flowers sunk in groups of three to seven in cavities on opposite sides of the joints. The species is found in Idaho and throughout the western United States in saline or alkaline soils, and marshy ground. The genus name is Latin (*sal*) for salt, and *cornu*, meaning horn and referring to its form and home.

Field Notes/Uses This and other *Salicornia* species are succulent and add a salty taste to salads. The young stems and branches can be pickled, but first must be boiled and then put into the pickling mixture.

Prickly Russian Thistle (*Salsola tragus*)

General Description This is not a true thistle (*Cirsium*), but a many branched annual with purplish striped stems up to three feet tall in a rounded form. The lower leaves are threadlike; the upper leaves are awl-like and spine-tipped. The plant may or may not be hairy. When mature, the whole plant becomes rigid, breaks off at ground level, and becomes one of the "tumbleweeds" that blows across the open plain. Flowers are solitary in the leaf axils and are subtended by spiny bracts. Russian thistle is common in open, disturbed habitats, particularly around agricultural areas at low elevations. It was introduced to the United States from Europe. Fortunately, Russian thistle is not an aggressive competitor and does not appear to replace native plant species. However, it is still considered a noxious weed because of its distributional pattern and spines.

Field Notes/Uses This unsavory looking plant is edible. The young parts of the plant may be boiled and eaten as a potherb or chopped raw into a salad. On older plants, clip the tender branch tips that are green. We find the taste of the plant greatly improves when cooked in butter and lemon. In Europe, the ashes of the plant were once used in the production of carbonate of soda known as Barilla (Clarke 1977).

Warning The older parts of the plants contain significant quantities of nitrates and oxalates and may be toxic if eaten in quantity.

Greasewood (*Sarcobatus vermiculatus*)

General Description This perennial native is a long-lived shrub with spreading, rigid branches that often bear spines. The leaves are linear, succulent, and pale green with entire margins. Some of the leaves may be opposite and some alternate, but all the leaves are shed in winter. The plants usually have both male and female flowers on the same plant (monoecious), but they can occur on separate plants (dioecious). Male flowers are catkin-like spikes on the ends of branches and the female flowers usually occur singly in the axial of leaves and form the fruits that are surrounded by a green membranous wing.

Field Notes/Uses Greasewood is used as wood for fuel and the sharpened spines were used for painting by Native Americans. Native Americans used the seeds and leaves, which have a salty taste, for food. The Hopi and other Native Americans use greasewood for fuel and for planting sticks. In Chaco Culture National Historical Park (New Mexico) greasewood was used for construction, especially of lintels, for fuel, being a preferred wood for Pueblo kiva fires.

Seeds, leaves, and new leaders are also consumed by a variety of small mammals. It is an important browse plant and is rated from good to useless forage for cattle, sheep, and big game animals in the winter and provides good cover and food for small mammals and birds. Sheep have been poisoned by rapidly consuming large amounts of new leader growth, which contains high levels of soluble oxalate. The numerous seeds are wind-dispersed and help to re-establish the plants after fire, although greasewood is only slightly harmed, if at all by fire, and will resprout.

Seepweed (Suaeda)

General Description These are annual or perennial herbs or small shrubs. Leaves are alternate, linear, square in cross-section or flattened. The flowers are perfect and occur in axils of small bracts. Suaeda is an Arabic name of antiquity. Two species *S. calceoliformis* (Pursh seepweed) and *S. moquinii* (Mojave seablite) occur in Idaho, with the former more widespread in the state.

Field Notes/Uses Identification of *Suaeda* specimens is achieved most successfully when based upon material containing flowers (for ovary shape) and mature calyces (for lobe shape) that contain the seeds. Because of the succulent nature of most specimens, fresh material may appear quite different than dried material, especially in the accentuation of calyx features when dry.

Plants of *Suaeda* are found in saline or alkaline wetlands or, occasionally, in upland habitats. Some species have been cultivated and eaten as a vegetable; seeds of some have been ground and eaten by Native Americans, and some species are used as a source for red or black dye.

St. John's-wort or Mangosteen Family - CLUSIACEAE

There are 40 genera and 1,000 species worldwide. The family is of little economic importance in North America. A few species are used as ornamentals.

St. John's-wort (*Hypericum*)

General Description The four perennial species that occur in Idaho have yellow flowers, and small, translucent glands on the leaves and petals. The species can be found in moist areas at various elevations.

Field Notes/Uses The largest and most widespread of the species is *Hypericum peforatum* (St. John's-wort). Weedon (1996) indicates that the leaves may be eaten fresh or may be dried and ground to a flour that can be used like acorn meal. However, only a small amount of the herbage should be consumed.

Despite its reputation as a weed, St. John's-wort may have much to offer humans as a medicinal plant. A number of clinical studies strongly suggest the plant may be effective in treating depression. Other laboratory studies reveal that St. John's-wort has at least two compounds, hypericin and pseudohypericin, which are active against retro-viruses. As such, it is being closely looked at in Acquired Immune Deficiency Syndrome (AIDS) research. The fresh flowers of St. John's-wort in tea, tincture, or olive oil, were once a popular domestic medicine for the treatment of external ulcers, wounds, sores, cuts, and bruises (Foster and Duke 1990). The ancient alleged magical properties of St. John's-wort were partly due to the fluorescent red pigment, hypericin, which oozes like blood from the crushed flowers. The red dye and extracts are used in cosmetics.

Caution Craighead et al. (1963) indicate that white-skinned animals feeding on these plants develop scabby sores and a skin itch. Apparently the plants contain photosensitive toxins and alkaloids. Therefore, ingestion is not recommended in large quantities.

Morning Glory Family - CONVOLVULACEAE

Members of this family are herbs, shrubs, or trees. In some species, there is milky latex present. The flowers are usually 5-merous with five united petals. There are approximately 50 genera and 1,400 to 1,700 species in this family, distributed in tropical and temperate regions. Nine genera are native to the United States. The family is of some economic importance because of the sweet potato (Ipomoea batadas), several weeds, and ornamentals.

False Bindweed (*Calystegia*)

General Description This is a fairly large genus of which many species have previously largely been referred to as Convolvulus. The various species also tend to integrate making identification difficult at times. The genus name is from the Greek, *kalux*, meaning cup, and *stegos*, meaning a covering. In Idaho, two species of *Calystegia* are known to occur: *C. hederacea* (Japanese false bindweed) and *C. sepium* (hedge false bindweed).

Field Notes/Uses The root of *C. sepium* was historically used as a purgative, and for treating jaundice and gall bladder ailments (Foster and Duke 1990). The root of a related species, *C. longipes* (Paiute morning glory), was boiled and a cup of the resultant brew was drunk for relief of gonorrhea. The dosage was taken every morning, noon, and evening until a cure was effected (Zigmond 1981).

Field Bindweed (*Convolvulus arvensis*)

General Description This is a perennial with trailing or twining stems. It spreads from a deep and brittle rhizome. The flowers are white or pinkish, funnel-shaped arising from the axils of the arrowhead-shaped leaves. This beautiful, but pernicious European weed is well established throughout North America and is frequently encountered on roadcuts and fields at low elevations. It is difficult to eradicate because of its deep rhizome and low growth. The genus name is Latin, *convolvere*, meaning to entwine.

Field Notes/Uses Native Americans used a cold leaf tea as a wash on spider bites. A tea from the flowers was used for fevers and wounds (Foster and Duke 1990). A tea was also made from the leaves and stems by the Kashaya and Pomo women to stop excessive menstruation (Strike 1994). In European folk use, the flower, leaf, and root teas were considered a laxative. The root is considered to be a strong purgative, cathartic, and diuretic. The powdered root stock was used as a laxative in ancient and modern China.

Dogwood Family - CORNACEAE

There are 12 genera and 100 species in the Dogwood Family including many ornamentals. A single genus, *Cornus*, occurs in the United States. Dogwoods are trees or shrubs, often with tiny flowers surrounded by petal-like bracts which resemble a single large flower. The leaves are opposite and simple.

Dogwood (*Cornus*)

General Description Dogwoods are shrubs or semi-woody perennials with simple leaves that are opposite or whorled. The flowers mature into red or white drupes. The four species in Idaho can be found in moist mountain and foothill forests, preferring partial shade, up to the subalpine zone.

Field Notes/Uses The fruits of *C. canadensis* (bunchberry dogwood) may be eaten raw or cooked. Since the berries are rather bland, we like to mix them with other more tasty berries. The unripe berries may cause stomachaches or act as a laxative. The chewed berries have been used as a poultice to treat local burns (Willard 1992). A cold and fever remedy can be made by boiling dried root or bark (the root is more potent). The feathered bark can be used as a toothbrush. Fresh bark is a cathartic. A leaf tea was used for aches and pains, kidney and lung ailments, coughs, fevers, and as eye wash (Foster and Duke 1990). Dogwood has earned a reputation as an anti-inflammatory and general analgesic due to the presence of cornine and other flavonoid compounds. Researchers are studying these properties as an anti-cancer agent. The current interest by pharmaceutical companies may stem from the fact that Native Americans used dogwood as an antidote for a variety of poisons.

The berries *of C. nuttallii* (Pacific dogwood) can be eaten raw or cooked. Strike (1994) suggests that the fruit contains enough protein, carbohydrate, and fat to sustain life when other food sources are not available. The inner bark of Pacific dogwood twigs can be scraped off and used as an additive to tobacco mixes for smoking. Pounded twigs can be used as a toothbrush. The wood for both species was used for bows and arrows, fishing hooks, and other implements. The bark was boiled and used to make a brown dye.

The berries of *C. sericea* (red-osier dogwood) were sometimes consumed by Native Americans. We have found them to be extremely bitter. In fact, in large quantities they may be toxic. The Blackfeet Indians used the bark of red-osier dogwood as a laxative. Other Native Americans smoked the inner bark as part of a ceremonial herb blend.

The last species, *C. unalaschkensis* (western cordilleran bunchberry) also was used by various Native Americans as food. The berries can be dried and used during winter, or in some instances, the berries were mashed, mixed with grease and eaten as a dessert.

CRASSULACEAE (Stonecrop Family)

Members of this family are succulent herbs or shrubs. There are 35 genera and over 1,500 species worldwide, of which nine genera are native to the United States. They are of no real economic importance except as ornamentals.

Ledge Stonecrop (*Rhodiola integrifolia*)

General Description Formerly known as *Sedum rosea* ssp. *integrifolium*. This plant has erect sturdy stems with terminal flat clusters of dark wine-colored flowers. Stems have many leaves, green to rosy pink in color and covered with white waxy powder. The leaves are flat, oval to spoon-shaped, becoming smaller on lower stem. Flowers have 4 or 5 petals and sepals, on erect stalks somewhat fleshy, dark reddish, occasionally yellow or pink. The plant grows in gravelly moist soils or among rocks at high elevations.

Field Notes/Uses The plant is edible raw or cooked, but slightly bitter (like some *Sedum* species). Alaska Natives chew the roots, spitting out the juice, to ease the discomfort of mouth sores, sore throats, and to bathe weary eyes. Parts of the plants are edible but can cause nausea if overeaten.

Stonecrop (*Sedum*)

General Description The eight species of stonecrop in Idaho are succulent, low growing perennial herbs. The Latin word *sedere* means "to sit," possibly referring to the tendency of many species growing low to the ground. Stonecrops are well-adapted to survival in shallow soil or on rocky outcroppings. The succulent leaves and stems have a waxy coating to help reduce water loss. The reddish color of the foliage in some species is enhanced by sunlight and occurs most often in plants in hot exposed sites.

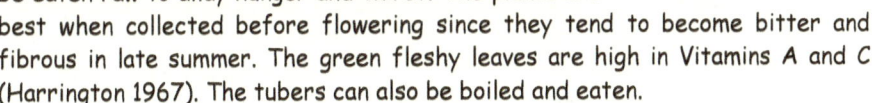

Field Notes/Uses The young leaves and stems of all species can be eaten as a salad or boiled as a potherb. We find them slightly tart and crisp - a wonderful addition to salads and trail snack. However, some species have emetic and cathartic properties, and can cause headaches. In an emergency, stonecrop can be eaten raw to allay hunger and thirst. The plants are best when collected before flowering since they tend to become bitter and fibrous in late summer. The green fleshy leaves are high in Vitamins A and C (Harrington 1967). The tubers can also be boiled and eaten.

Sedum has also been reported as being slightly astringent and mucilaginous. It is valuable in the treatment of wounds, ulcers, lung disorders, and diarrhea (Willard 1992). As a field remedy for minor burns, insect bites, and other skin irritations, just squeeze the juice onto the affected area. Decoctions of the plant were also used for sore throats and colds, and as an eye wash (Pojar and MacKinnon 1994).

Crossoma Family - CROSSOSOMATACEAE

Members of this family are shrubs with simple, entire, and coriaceous leaves. The flowers are white or sometimes rose-tinged, and the fruit (follicle) is green at maturity. There are 3 genera and about 8 species in the western United States and Mexico.

Spiny Greasewood (*Glossopetalon spinescens*)

General Description This is a densely branched, deciduous shrub that may grow up to 10 feet tall. The spreading, spine-tipped branches have green bark and are grooved lengthwise. The alternate, grayish green leaves are entire, usually broadly lance-shaped. The white petals are $\frac{1}{4}$ to $\frac{1}{2}$ inch long, and there are 5-8 stamens. The leathery capsule is grooved lengthwise and about 1/8 inch long.

Field Notes/Uses This is an unusual, uncommon, spiny shrub found in most western states but most often in scattered or rare populations. The narrow petals of its white flowers twist and curl and some petals fall before others, giving the plant an unkept appearance. The botanist Asa Gray named this plant in 1853 from a specimen collected by Charles Wright "in a mountain ravine near Frontera, New Mexico (now Texas)" in 1852. The Shoshoni made a decoction of the shrub that was used for tuberculosis.

Cucumber or Gourd Family - CUCURBITACEAE

Members of this family are annual or perennial herbs that are climbing or prostrate, with spirally coiled tendrils. The leaves are alternate, often palmately lobed. The fruit is a berry (often referred to as a pepo) with a leathery or hard exocarp. There are approximately 100 genera and 850 species distributed in the warmer regions of the Old and New World. Fourteen genera are native to the United States. The family is economically important as a source of many food plants and ornamentals.

Wild Cucumber (*Echinocystis lobata*)

General Description Wild cucumber is a high climbing vine with angular, grooved stems. The alternate leaves are palmately lobed. The flowers are greenish-white and develop into egg-shaped, prickly fruits. Widely distributed across North America, it is usually found at low elevations.

Field Notes/Uses Native Americans used the bitter root tea as a tonic for stomach troubles, kidney ailments, rheumatism, chills, and fevers. The roots were also pulverized and then poulticed for headaches (Foster and Duke 1990).

Dodder Family - CUSCUTACEAE

Members of the Dodder family are leafless, rootless, parasitic herbs that lack chlorophyll. The stems are thread-like and often yellowish in color. The small flowers have 4 or 5 distinct sepals, and 4 or 5 united petals. The fruit is a dry or fleshy globose capsule. The family has one genus (*Cuscuta*) with approximately 170 species. The genus is native to the United States and may cause great losses to crop plants. The family was once included in the Convolvulaceae (Morning Glory family).

Dodder (*Cuscuta*)

General Description Dodders are leafless, twining perennials with slender stems that colored pink, whitish, or yellowish, never green. Both the leaves and pink to white flowers are highly reduced. Dodder can normally be identified only with a microscope or hand lens. The many species of *Cuscuta* parasitize different flowering plant hosts at low elevations.

Field Notes/Uses Dodders have a very unique life cycle. The small seeds usually germinate in the soil and produce slender stems without seed leaves (cotyledons). Unless the slowly rotating plant encounters a host plant within a short period of time, the dodder seedling will wither and die. However, if the seedling encounters the living stem of a susceptible host plant, the dodder will twine around it and at certain points develop suckers that penetrate the tissue of the host. Nutrition is received through these suckers. Dodder then losses all contact with the soil. After a period of growth, small flowers develop and large amounts of seeds are produced to start the process all over again (Franklin and Mulligan 1987).

Dodder was often called "love vine" and "vegetable spaghetti" by some Native Americans, but they are generally not considered to be edible and may cause digestive upset. However, the seeds of *C. californica* (chaparral dodder) were parched and eaten by Native Americans in California (Strike 1994).

Chaparral dodder, when brewed as a tea, was considered to be an antidote for black widow bites. However, only the dodder from *Eriogonum fasciculatum* (California buckwheat) was used for this purpose (Strike 1994). Other Native Americans chewed a mass of dodder and stuffed it in their nose or pulverized the plant and sniffed the powder to stop nosebleeds.

Dodder stems were used by the Cherokees as a poultice for bruises. In China, the stems of some species of dodder are used in lotions for inflamed eyes. Moore (1979) indicates that a rounded teaspoon of the chopped dodder is a good laxative-cathartic. In smaller quantities, and drunk every few hours, it is said that it will aid in spleen inflammations, lymph node swellings, and "liver torpor." Additionally, handfuls of dodder can be gathered and used as scouring pads for cleaning.

Teasel Family - DIPSACACEAE

These are mostly herbaceous plants with opposite leaves. There are approximately 10 genera and 270 species found mostly in the Old World. None are native to the United States, although *Dipsacus* is widely naturalized and has become quite weedy.

Fuller's Teasel (*Dipsacus fullonum*)

General Description Fuller's teasel is a stout, prickly tap-rooted plant up to six feet tall. The leaves are opposite and lance-shaped and distinctly prickly on the lower surface of the midrib. The small, bluish-purple flowers occur in a large, terminal flower head that is egg-shaped and armed with numerous, sharp-pointed bracts. This non-native plant from Europe is found throughout North America on disturbed soils with appreciable water holding capacity. The plant has a tendency to be opportunistic, and displaces desirable native plants. The genus name is from Greek *dipsa*, which means thirst, and refers to the accumulation of water in the cup-like bases of the joined leaves. It is not uncommon to find insects and other invertebrates living in the water found in the leaves.

Field Notes/Uses The dried flower spike of teasel looks like a caged "thistle" and is sometimes mistaken for thistle (*Cirsium*) because of its "thorny"

character and bluish flowers. However, thistle is a member of the Sunflower family (Asteraceae) and has alternate, not opposite leaves.

While there are no documented edible or medicinal uses for the plant, they are sought after for use in dry flower arrangements. The dried inflorescence is often sprayed with some interesting colors and sold in craft stores.

Cloth cleaners use the dried flower heads to remove the nap from wool or cloth after beating and cleaning. Apparently, the teasel heads perform the task so well that man-made tools have not replaced them.

Sundew Family - DROSERACEAE

Members of the Sundew family are insectivorous herbs growing in acidic bogs. The leaves are covered with sticky glandular hairs on which insects become trapped. There are approximately four genera and 100 species, all of which grow in very nutrient poor soil conditions. *Drosera*, a native of the United States, is cosmopolitan in its distribution, whereas the other genera are monotypic and restricted in their distribution. The family is of no economic importance, except that several species are cultivated as novelties because of their insectivorous habit.

Sundew (*Drosera*)

General Description The three species in Idaho, *D. anglica* (English sundew), *D. intermedia* (spoonleaf sundew), and *D. rotundifolia* (roundleaf sundew), are perennials with basal rosettes of leaves covered with sticky glands that trap and digest small insects. They are usually found in bogs in association with sphagnum moss. The occurrence of *D. intermedia* in Idaho is questionable given the species is an eastern species.

Field Notes/Uses The juice of some *Drosera* species has been used to curdle milk. In the Mediterranean region, sundews are mixed with brandy, raisins, and sugars, and allowed to ferment into a drink called Rossolis. However, the main use of *Drosera* has been medicinal.

The leaves of *D. rotundifolia* apparently have antispasmodic and expectorant properties. It has been used in the treatment of whooping cough, bronchitis, asthma, and other respiratory problems. To relieve a bad cough, the leaves were tinctured or made into a tea, and sipped throughout the day. The plant also contains an antibiotic substance that, in pure

Tinctures

Here is one general procedure for making tinctures. First, place the fresh plant in a glass jar and cover with 80 proof brandy or vodka. Keep the jar in a warm dark place for about two weeks, shaking it daily. After two weeks, strain the herbs through muslin or two layers of cheesecloth. Squeeze well to extract as much fluid as possible. Discard the herbs and bottle the tincture. The dosage varies depending on what the herb is to be used for. For sundew, 3-6 drops in a cup of water is recommended.

Vinegar can be used in place of alcohol, but has a shorter shelf life (1-2 years) than alcohol tinctures (30-40 years). Vinegar tinctures are often used for babies, form, is effective against streptococcus, staphylococcus, and pneumococcus (Lust 1974).

Schofield (1989) reports that sundew is helpful in a tinctured form for nausea associated with seasickness. Apparently, some drowsiness was associated with the tincture. Sundew juice has been used for removing warts, and has been blended with milk to lighten freckles.

Caution While there are no reports of human poisoning from ingestion of sundew, they contain corrosive, irritating substances and should only be used in small doses. Cats have been poisoned from daily doses of the plants.

Oleaster Family - ELAEAGNACEAE

Plants in this family are shrubs or trees with alternate and silvery-gray leaves. The leaves are alternate or opposite, generally deciduous. The fruits are drupe- or berry-like in appearance. There are three genera and approximately 45 species in this family distributed in North America, south Europe, Asia, and eastern Australia.

Silverberry (*Elaeagnus*)

General Description The two species of *Elaeagnus* in Idaho are shrubs or trees with silvery-gray leaves. The egg-shaped fruits are also silvery-gray. Introduced from Asia, *E. angustifolia* (Russian olive) is used as a windbreak.

Field Notes/Uses The native species, *E. commutata* (silverberry) is found along gravel river bars and water courses. Silverberries may be eaten raw, but are dry and mealy and not highly regarded. Most often, Native Americans fried them in moose fat or some other grease. They should be considered as an emergency food. The flowers could be used in making perfumes and other cosmetics.

Russet Buffaloberry (*Shepherdia canadensis*)

General Description Two species occur in Idaho, but it is russet buffaloberry that is more commonly encountered in the state. It is a

"Indian Ice Cream"

To make Indian ice cream, Native Americans placed a small number of berries into a bowl with a little water, then used a special stick with some grass tied on one end to beat the fruit. The result was foamy concoction. In recent times, sugar was added to improve the taste. Care must also be taken in picking and preparing the berries so that they do not come in contact with oil or grease of any kind, or they will not whip. Indian ice cream is still served in many households, especially at parties and family gatherings low spreading shrub with opposite leaves, each with a dark green upper surface and lighter underside that is covered with tiny, brown scales. The inconspicuous flowers are yellow-green; the fruits range in color from yellow to bright red. Buffaloberry is common in forested environments, from the montane to subalpine/lodgepole pine forests. It is also found in recently burned areas. The genus is named after John Shepherd (1764-1836), curator of the Liverpool Botanic Garden.

Field Notes/Uses Another common name for this species is "soapberry." The berries contain a significant amount of saponin which not only gives the plant its bitter taste, but also whips up into a frothy mass called "Indian Ice Cream."

Native Americans used the berries of these plants extensively, both fresh and dried. The berries of this species are at first pleasant, then the soap-like bitterness prevails. We enjoy cooking them with sweeter tasting berries such as thimbleberries (*Rubus*) and serviceberries (*Amelanchier*), and a large amount of sugar. We found the berries to be somewhat unattractive for general use, but a valuable consideration in emergencies. The berries taste better after a few good frosts during the fall. They can also be used in the making of pemmican or jelly. Dried into cakes, the berries can be stored for winter. The fruits of *S. argentea* (silver buffaloberry), the other species in Idaho, are also edible and do not contain as much saponin as *S. canadensis*.

Infusions of the stems and leaves were drunk as a tonic beverage (Schofield 1989). The berries can be crushed and made into a tea for use as a liquid soap. Native Americans used the tea to relieve constipation. Hart (1976) indicates that the Flathead and Kootenai Indians made solutions from the bark of buffaloberry for eye troubles.

Heath Family - ERICACEAE

There are about 50 genera and 2,500 species in the Heath Family, the majority occurring in acidic soils. In the United States approximately 25 genera are indigenous. Economic products provided by this family include food plants, oil of wintergreen, and many ornamentals. Some herbaceous members are mycotrophic, that is, they depend on fungi for nutrient uptake and lack chlorophyll. The list of genera and species here includes the parasitic and hemiparasitic species of Monotropaceae and Pyrolaceae.

Sugarstick (*Allotropa virgata*)

General Description The white and pink striped stem up to 16 inches tall renders candystick a distinctive plant in the wild. When picked and dried, it fades to a dull black. The whitish leaves taper gradually to a point that curves outward. Though lacking a corolla, the spike-like raceme is still colorful due to pink, white or brownish sepals. The flower stalk is distinct in its white and red or maroon stripes from which the common name sugarstick follows.

Field Notes/Uses Sugarstick is an herbaceous perennial with a geographic distribution in the western United States, California, and Nevada north to Washington and east to Montana, where it is commonly encountered. Sugarsticks occur above ground as a cluster of flowering stalks. This is a Forest Service Sensitive species in the Intermountain Region.

This plant lacks both chlorophyll and leaves, and uses complex relationships with a fungal intermediary (e.g., *Tricholoma matsutake*) to obtain energy from the roots of other plants. These mushrooms are big business, with the majority harvested being exported to Japan. These edible mushrooms are prized in Japan, both for their flavor and meaning. To this day they're still given as important gifts, meant to symbolize fertility and happiness.

Bog Rosemary (*Andromeda polifolia*)

General Description Bog rosemary is named for its narrow, leathery leaves with white undersides, which resemble rosemary spice (Rosmarinus officinalis). This low-growing shrub inhabits mossy regions of bogs. The delicate nodding pink flowers decorate northern bogs. The genus was named by Carl Linnaeus who observed it Lapland and compared the plant to Andromeda from Greek mythology. The specific epithet *polifolia* means "grey-leaved."

Field Notes/Uses Bog rosemary contains andromedotoxin. Domestic browsing animals usually reject bog rosemary, but the plant has killed domestic sheep. The heather vole feeds on leaves and fruits of bog rosemary.

The leaves are used to make a tea. Bog rosemary contains grayanotoxin, which when ingested lowers blood pressure, and may cause respiratory problems, dizziness, vomiting, or diarrhea.

Kinnikinnic (*Arctostaphylos uva-ursi*)

General Description This is a low, mat-forming shrub with reddish to brown stems that root at the nodes. The spoon- to lance-shaped leaves are leathery. The flowers are urn-shaped and the bright red berries often persist through the winter. It is found in open areas with dry to well-drained soils from low to high elevations. The genus name means "bear grape" and refers to the fondness shown by bears for the fruits of these shrubs, many of which are known as Bearberry.

Field Notes/Uses While only kinnikinnic occurs in Idaho, the berries of all *Arctostaphylos* (manzanita) are edible. They may be eaten raw, and it is suggested that they not be eaten in large quantities since they may be hard to digest. Constipation or indigestion are common maladies of eating too much. The berries can also be stewed, or dried and ground into meal and cooked as mush. A cider can also be made from the berries. The seeds alone can be collected and ground into meal too.

Manzanita Cider

To make manzanita cider, simply crush the ripe or green berries, in a container, and then pour an equal volume of scalding water over them. After the mixture has cooled and the solids have settled, decant the liquid and drink. The cider will be a little dry to the taste.

Leaves and bark when dried can be smoked as a tobacco substitute. Harrington (1967) indicates that the general effect of some Native American tribes smoking the leaves of kinnikinnic was intoxication due to its narcotic content. Leaves are an astringent due to the tannic acid and have been used to tan hides. The leaves can also be chewed to stimulate saliva, particularly when one is thirsty.

The leaves can also be boiled in water, allowed to cool, and the decoction applied to stop the itching and spread of poison ivy (Angier 1978). The internal consumption of the leaf tea often results in urine becoming alkaline and bright green. This is caused by the urinary antiseptic hydroquinolone, and it is relatively harmless. These hydroquinones (particularly arbutin) are strongly antibacterial and are effective against *Klebsiella* and *E. coli*, which are often associated with urinary infections.

Western Moss Heather (*Cassiope mertensiana*)

General Description This is a low-growing, creeping and evergreen shrub of high mountain environments. It has small, scale-like leaves and white, bell-shaped flowers. The delicate, white, bell-shaped corollas are set off by reddish sepals.

Field Notes/Uses Thompson Indians used a decoction of western moss heather taken over a period of time for tuberculosis and spitting up blood

Pipsissewa (*Chimaphila*)

General Description Two species of *Chimaphila* can be found in Idaho, *C. menziesii* (little prince's pine) and *C. umbellata* (pipsissewa). They are short evergreen semi-shrubs (woody only at the base) that originate from a long creeping rootstock. The leaves are whorled and leathery. Both species are common in forests up to the subalpine zone.

Field Notes/Uses The roots and leaves of pipsissewa may be boiled, and the liquid cooled for a refreshing drink that is high in Vitamin C. The leaves may also be nibbled raw, but because of their astringency and tough texture we found them unappealing.

Pipsissewa was an important herb to Native Americans for treating rheumatism. A tea from the leaves was used for the purpose to treat rheumatism and kidney problems (Foster and Duke 1990). The plant contains quinone glycosides, such as that found in *Arctostaphylos*, but is less astringent and more a diuretic, making it better for long-term use. The plant was also mixed with tobacco for smoking. Pipsissewa produces a natural antibiotic that can be used by humans (Willard 1992, Moore 1979). Hot infusions of pipsissewa can be taken to induce perspiration in the treatment of typhus, and the berries can be eaten for stomach disorders (Moore 1979).

Pipsissewa is a "secret ingredient" in certain popular soft drinks. In the Northwest, these plants, as well as certain species of *Pyrola* are under commercial harvesting pressure and may be slowly disappearing.

Snowberry (*Gaultheria*)

General Description The three species of *Gaultheria* in Idaho are dwarf evergreen shrubs that form small mats with leaves that are broadly egg-shaped to elliptical. The bell-shaped flowers are white to pink and the berries are red. Snowberries can be found in moist areas at mid elevations.

Field Notes/Uses The small, red fruits of *G. humifusa* (alpine spicywintergreen) are edible raw or cooked, and can be made into jams, wines, or pies. The young tender leaves are suitable as greens and have a wintergreen flavor. The fresh leaves of *G. hispidula* (creeping snowberry) can be used to make a tea, and the berries are also edible.

In Native American medicine, the plants were used for treating aches and pains and to help with breathing while hunting or carrying heavy loads. The leaves of these species yield oil upon steam distillation. This "oil of wintergreen" (methyl salicylate) is a folk remedy for body aches and pains, and is known for its astringent, diuretic, and stimulant properties.

Warning The wintergreen flavor in the plants is due to the presence of oil of wintergreen, which, if taken in excess can be toxic, especially to children. In small amounts, such as in wintergreen tea, there is little danger. Children who are allergic to aspirin (a related drug) should not eat the plant or berries, or even handle the plant.

Alpine Laurel (*Kalmia microphylla*)

General Description These are branched, evergreen shrubs that spread by short rhizomes and layering. The leaves are opposite, lance-shaped to elliptical and have in-rolled margins. The flowers are rose-colored and bowl-shaped. The species occurs in the mid- to upper elevations, usually in moist to wet, acidic soils, often along creeks.

Field Notes/Uses The toxicity of *Kalmia* is legendary. Some Native Americans used it as a suicide plant. Game birds and livestock may be poisonous to eat if they have ingested the leaves. According to Peter Kalm (1715-1779), after whom the genus is named;

> "...sheep are especially susceptible, while deer are unharmed. Though the flesh of affected animals is apparently not contaminated, the intestines will cause poisoning if fed to dogs so that they become quite stupid and as it were intoxicated and often fall so sick that they seem to be at the point of death."

All *Kalmia* species should be considered poisonous. They contain andromedotoxin, which causes a slow pulse, low blood pressure, lack of coordination, convulsions, progressive paralysis, and death. The honey made by bees from these plants is also poisonous.

Labrador Tea (*Ledum*)

General Description Two species of Labrador tea can be found in Idaho. They are evergreen shrubs, with short, fine hairs and glands on young branches and lower leaf surfaces. The leaves are elliptical to egg-shaped and are clustered near the stem tips, giving them a whorled effect. The white flowers are in rounded clusters at the tip of the stem. All parts of the plant smell like turpentine when crushed. Look for the plants in the mid- to upper subalpine elevations, particularly in permanently wet or moist, acidic soils.

Field Notes/Uses Labrador tea contains ledol, which is a narcotic toxin that causes drowsiness, delirium, cramps, paralysis, heart palpations, and even death if taken in excess. Prolonged cooking extracts large doses of ledol. Otherwise the tea is slightly laxative. Infusions are recommended for camper's distress (e.g., constipation). The leaves are astringent and useful in facial creams.

Andromedotoxin is also found in the leaves of these plants (see Kalmia above), and therefore they should be considered poisonous. The leaves of L. groenlandicum (bog Labrador tea) make a mild but agreeable tea when steeped in hot water (Peterson 1977). Willard (1992), Densmore (1974), and Foster and Duke (1990) indicate that the tea was used for colds, rheumatism, scurvy, and stomach ailments, but is not recommended. A strong decoction of the leaves was used as a wash to get rid of lice. As an insect repellent, it was said to be quite effective against mosquitoes (Willard 1992). According to Elias and Dykeman (1982);

> "... steep 1 tablespoon dried leaves in cup of boiling hot water for 10 minutes. Do not boil water with leaves in it, as it may release the harmful alkaloids. Serve hot or cold."

Ledum glandulosum (western Labrador tea), another species that can be encountered in Idaho has a common name of "Trapper's Tea." The name apparently originated not because trappers drank it, but because they boiled their traps in it to de-scent them.

Warning Many plants in this family contain a poisonous compound called andromedotoxin. If consumed in large concentrations this could be harmful, causing vomiting, illness, and even death. In comparison, *Ledum* evidently has less andromedotoxin than other related plants such as *Kalmia* and *Rhododendrons*; nevertheless, *Ledum* should be used only in moderation and in relatively dilute infusions. It is also very important to not confuse *Ledum* with the more toxic *Kalmia* which grows in similar habitats.

Rusty Menziesia (*Menziesia ferruginea*)

General Description This is a deciduous shrub with shredding grey-brown bark. The leaves are crowded on the stem giving a whorled effect. The plant has a skunk-like odor when crushed. The cinnamon-pink flowers are small and urn-shaped. It can be found in wooded areas, usually on north and east facing slopes from the mid- to upper elevations.

Field Notes/Uses The twigs and leaves of this species were used by some Native Americans to make a tea, but the plant contains some poisonous alkaloids. Additionally, a fungus (*Exobasidium* sp. affin. *vaccinii*) growing on the leaves of this shrub was eaten by some aboriginal peoples (Pojar and MacKinnon 1994).

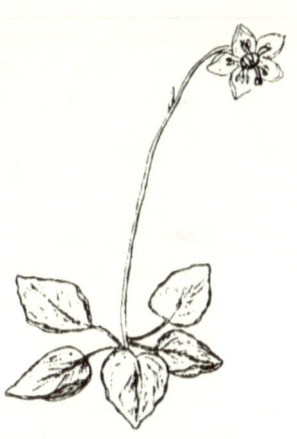

Single Delight (*Moneses uniflora*)

General Description The leaves of this plant are basal, thin, ovate in shape, and sharply serrulate. The flowers are solitary and the petals are white to pink in color. The anthers are 2-horned and the stigma 5-lobed. Not a common plant.

Field Notes/Uses A poultice of the leaves was applied to draw out pus from boils. An infusion of the dried plants was used for coughs and colds, and the plants were chewed for sore throats. A poultice of the chewed or pounded plant was applied to pains. The fruit was supposedly used as food by Montana Indians – though unknown how (Hart 1996).

Pine Sap, Indian Pipe (*Monotropa*)

General Description Two species are found in Idaho: *M. hypopithys* (pinesap) and *M. uniflora* (IndianpPipe). *Monotropa uniflora* has waxy-white stems that blacken with age or drying. It is usually found in thick humus in the deep shade of coniferous forests.

Field Notes/Uses Indian pipe is said to be edible raw or cooked (Kirk 1975). The plant juice mixed with water was used as an eye lotion (Coffey 1993). Settlers used the fresh juice for a wide range of eye ailments and the dried plant was used in place of opium to relieve pain and induce sleep (Foster and Duke 1990). *M. hypopithys* is a fleshy yellowish to pinkish saprophyte that dries black. In occurs in forests at the middle elevations, and is edible raw or cooked (Kirk 1975).

Sidebells Wintergreen (*Orthilia secunda*)

General Description This species was once in the genus Pyrola (*P. secunda*). Sidebells wintergreen has stems up 6 inches tall clustered from a long, thin, branched rhizome. Several broadly lance-shaped to elliptical, olive green leaves and more than twice this number of small bracts are arranged on the lower part of each creeping and ascending stem. The small white flowers with petals that fall shortly after opening are distinctively arranged on 1 side of the stalk (hence name *secunda*, for one-sided).

Field Notes/Uses A strong decoction of the root was used as eyewash.

Mountainheath (*Phyllodoce*)

General Description: There are two species in Idaho. These are the common heathers of subalpine meadows, although not the same as the classic heather of Europe. These low shrubs grow as spreading mats, with many short , upright stems. The evergreen leaves have glandular margins recurved and the underside appearing grooved. The flowers occur at the tops of the stems and usually nod on glandular stalks. The fruit is a roundish, dry capsule.

Field Notes/Uses: Pink mountainheath (*P. empetriformis*) was used by the Thompson Indians (in southwestern British Columbia, Canada) as a tuberculosis remedy. Apparently, a decoction of the plant was taken over a period of time for tuberculosis and spitting up blood.

Woodland Pinedrops (*Pterospora andromedea*)

General Description Woodland pinedrops is a brownish-red plant with sticky stems up to 3 feet tall with pale-yellow flowers. Found in deep humus of coniferous forests, it is usually associated with ponderosa pine (*Pinus ponderosa*).

Field Notes/Uses Foster and Duke (1990) indicate that Native Americans used a cold tea made from the pounded stems and fruits to treat bleeding from the lungs. As a dry powder, the plant was used as a snuff for nosebleeds.

Wintergreen (*Pyrola*)

General Description Five species of *Pyrola* occur in Idaho. In general, they are low, smooth perennial herbs with shiny, leathery leaves that are clustered at the base. The flowers are waxy and nodding. *Pyrola* stems from pyrus for pear, probably since the leaves of many species resemble pear leaves.

Field Notes/Uses A tea made from the whole plant was used to treat epileptic seizures in babies (Foster and Duke 1990). A leaf tea was gargled for sore throats, and canker sores, while a tea from the root was a tonic. A poultice from the mashed leaves was used for tumors, sores, cuts, and to relieve the itch of insect bites (Foster and Duke 1990). The plant is also an excellent astringent and disinfectant for urinary tract infections (Tilford 1993). The plants contain ursolic acid and the glycosides arbutin and ericolin, which were used in the treatment of kidney problems and skin eruptions (Schofield 1989).

Pyrola is also used as an ingredient in popular soft drinks. It is said to be an excellent substitute for *Chimaphila umbellata* (pipsissewa). In some areas *Pyrola* may be exploited (over harvested) for commercial purposes (Tilford 1997).

Cascade Azalea (*Rhododendron albiflorum*)

General Description Cascade azalea is a deciduous shrub with coarse reddish hairs on young twigs, lower leaf surfaces, and flowers. The leaves are elliptical to lance-shaped, and are clustered at the stem tip appearing to be whorled. The white flowers of the plant blossom in late July and can be found along streambanks and moist slopes from mid- to subalpine elevations.

Field Notes/Uses The leaves may contain andromedotoxin and should be considered poisonous (see *Kalmia*). Pojar and MacKinnon (1994) report that the leaves of white rhododendron were used as a tea by the Okanogan, and the Skokomish used the buds boiled in water as a cold and sore throat medicine.

Blueberry, Dwarf Huckleberry (*Vaccinium*)

General Description The eight species of *Vaccinium* in Idaho are small to mid-sized shrubs with deciduous leaves. The twigs are often angled. The small flowers are urn-shaped, and fruits many seeded berries. They can be found on well-drained sites, from wet meadows and around lakes up to the timberline.

Field Notes/Uses *Vaccinium* berries can be eaten raw or be dried in the form of cakes for future use. The various species we have sampled range in taste from sweet to tart. Hybridization between the species is known to occur, but the fruits are still edible. The berries have also been used as fish bait since they look very similar to salmon eggs. The leaves can be dried to make a tea. The leaves and berries are high in Vitamin C.

Spurge Family - EUPHORBIACEAE

The Spurge family has about 290 genera and 7,500 species distributed worldwide. Among the valuable products of the family are rubber, castor and tung oils, and tapioca. Most members are poisonous, and have milky sap that will irritate eyes and mouth.

Turkey Mullein (*Croton setigerus*)

General Description Turkey mullein is a grayish-green annual with a musky smell. It is usually found in dry, often rocky areas at low elevations.

Field Notes/Uses The herbage of the plant is probably poisonous (Muenscher 1940). Strike (1994) indicates that the Costanoan Indians used a decoction of the root for dysentery, while the Kawaiisu used the decoction, internally and/or externally, to relieve headaches and rheumatic pains.

Sweet (1976) says that the plant contains a narcotic and the foliage was used by some aboriginal peoples to stupefy fish and poison their arrow points. It has been suggested that the stellate hairs get into the fish's gills and hold them open, so in time the fish would drown (Murphey 1990). Moore (1989) on the other hand, indicates that it may be diterpenes that are the main cause of the plant's effects on fish. Fresh leaves were bruised and applied as a counterirritant poultice for internal pain and asthma (Balls 1962).

Sandmat/Spurge (*Chamaesyce/Euphorbia*)

General Description There are 4 species of *Chamaesyce* and seven species of *Euphorbia* in Idaho. They are annual or perennial herbs with milky juice. The flowers are borne in a complex structure called a cyathium. This cup-like structure contains several male flowers and a single female flower. They are found in disturbed habitats in low elevations.

Field Notes/Uses *Euphorbia* contains toxic principles that will cause severe poisoning if ingested in quantity. Most species contain carcinogenic, highly irritant, diterpene esters and are strong purgatives. The white sap can cause skin irritations and blisters (Willard 1992). Bean and Saubel (1972) report that Native Americans in California used both the native and introduced species as a medicine for reducing fever and as a cure for chicken pox and smallpox. The plant was boiled and the afflicted person was bathed in the decoction.

Euphorbia name is from the Greek "euphorbion," a plant named after *Euphorbos*, a celebrated Greek physician of the 1st century B.C. *Chamaesyce* is from the Greek for "creeping fig." *Chamaesyce* is often pronounced – "camee sigh see." The milky sap of these and other spurges acts as a deterrent for herbivores as well as a wound healer. Like other euphorbs, the sap contains diterpene esters and caustic to the skin and can be painful if it came in contact with the eyes.

Pea or Legume Family - FABACEAE

The Pea family is one of the largest plant families in the world. There are approximately 600 genera and 13,000 species worldwide. Next to the grass family, which produces all our grains and cereals, the pea family is the second most economically important group of plants in the world. The beans and peas that we eat for dinner, as well as the traditional peanuts at baseball games are found in this family. But before taking a bite of the next legume you see, be aware that the family also contains a number of highly toxic members. The various species of locoweeds and milkvetches (*Oxytropis* and *Astragalus*) have caused much loss of livestock.

The pea flower is referred to as "papilionaceous," and means butterfly-like. The flowers are bilaterally symmetrical, consisting of 5 petals. The largest upper petal is called the "banner," the two lateral ones are the "wings", and the two lowest ones are fused at the lower margins to form a boat-like structure called the "keel."

Milkvetch (*Astragalus*)

General Description Many species of milkvetch occur in Idaho. They are perennial herbs with odd-pinnate leaves that have leafy stipules. This is a difficult genus of perhaps 1600 species, making it the largest genus in the Pea family. The name comes from the ancient Greek name for a plant in the pea family.

Field Notes/Uses Although the roots, pods, and peas of some species were reported to be eaten by American Indians (Craighead et al. 1963), this genus is not recommended for consumption. All milkvetches produce either a toxic alkaloid substance or accumulate selenium from the soil or both. Selenium poisoning of livestock has the following characteristics - lethargy, diarrhea, loss of hair, breakage at the base of hoof, excessive urination, difficulty breathing, rapid and weak pulse, and coma. Death usually results from the failure of lungs and heart.

An interesting side benefit has developed from discovery of plants that grow only in selenium soils. Scientists can use the plants to map areas high in selenium for the purpose of mining the valuable element. Some Astragalus species also provide good indicators of uranium ore and copper-molybdenum deposits.

American Licorice (*Glycyrrhiza lepidota*)

General Description This is a perennial herb growing up to 3 feet tall. The flowers are greenish-white in dense racemes. Mature fruits are a conspicuous pod up one-half inch long, and densely covered with hooked spines. Usually found in moist, sandy soils, river banks at the lower elevations.

Field Notes/Uses The plant contains glycyrrhizin, sugar and other chemicals used in medicine as a mild laxative, a demulcent, and a flavoring to mask the taste of other drugs. It is also used in confections, root beer, and chewing tobacco. Licorice root has been used in the treatment of asthma, stomach ulcers, bronchitis, and urinary tract disorders (Tilford 1997). The plants were chewed by Natives and used as a flavoring.

Warning Continual use of this plant in large doses may cause water retention and elevated blood pressure.

Sweetvetch (*Hedysarum*)

General Description The three species of Hedysarum in Idaho are *H. boreale* (northern sweetvetch), H. sulphurescens (white sweetvetch), and *H. occidentalis* (western sweetvetch). They are perennial herbs that are sparsely branched and slightly hairy. The stipules are usually brown, large and sheath-like on the lower stem, and are narrow and pointed on the upper stem. The typical pea-like flower has a much longer keel than wings and banner. The species occur in open forest, meadow, and rocky ledges from the mid- to higher elevations. The botanical name is from the Greek hedys meaning sweet, and aroma for smell. The flowers are quite fragrant.

Field Notes/Uses Harvest *Hedysarums* only if you can positively identify the species you are gathering. Unlike other genera where some species are better tasting than others, but all are still harmless, the *Hedysarums* do not follow that pattern in terms of safety for eating. A person wishing to harvest these plants must make a special effort to notice the subtle variations in leaf and flower structure between species. A misidentification can easily lead to some really upset stomachs. Fortunately, all three species that are found in Idaho have sweet tasting edible roots. Early settlers and Native Americans used the roots of *H. boreale* as a licorice substitute.

Caution It is important to correctly identify any *Hedysarum* to species that is intended to be consumed. *Hedysarum* is also sometimes commonly confused with *Astragalus*, but can be easily distinguished when pods are present. *Hedysarum* is distinguished from *Astragalus* by the presence of a loment, and the keel is longer than or equaling the wings and banner.

Sweetpea (*Lathyrus*)

General Description The nine species of sweetpeas are vines, climbing or supporting themselves on other vegetation. The plants have tendrils at the ends of their leaves. Sweetpeas are found in a variety of habitats from the foothills up to the subalpine.

Field Notes/Uses Some species of *Lathyrus* have a history of poisoning humans. Kirk (1975) indicates that an exclusive diet of some species from 10 to 30 days can bring on partial or total paralysis, and Willard (1992) suggests avoiding these plants entirely. Strike (1994) says that the greens and raw seeds of *Lathyrus* in California were eaten by Native Americans. Some of the seeds were parched and made into pinole and was stored for winter use. Weedon (1996) says that the fruit of many species are edible in small amounts, but may cause paralysis and several secondary disorders if eaten in large quantities over time.

Caution It is best to assume that these plants are poisonous if ingested.

Trefoil (*Lotus*)

General Description The eight species of Lotus are annual or perennial herbs with pinnately compound leaves. The flowers are pea-shaped, yellow or white, often tinged with reddish or purple. They occur in many habitats at various elevations.

Field Notes/Uses Many of the species are presumed to be poisonous and should be avoided.

Lupine (*Lupinus*)

General Description The many species of Lupine in Idaho are showy perennial or annual herbs with palmately compound leaves. Flowers are blue, violet, rarely white, rose in elongated narrow inflorescence. The pods are flattened and usually hairy. They are found on open slopes and meadows up into the alpine zone. Latin name of *Lupinus* comes lupus, meaning wolf, alluding to the belief that this plant wolfed nutrients and caused poor soil conditions. To the contrary, lupines are nitrogen fixers that greatly improve soil conditions.

Field Notes/Uses The pea-like seeds have been wrongly recommended by some authors of edible plant books as a substitute for peas. Lupines posses

many complex alkaloids and should be considered poisonous. Records do indicate, however, that some species have been safely consumed. For example, Weedon (1996) and Scully (1970) indicate that the young leaves and unopened flowers were steamed and eaten with soup by some Native American tribes. But because of hybridization, the edible species can concentrate toxic alkaloids that could result in an unhealthy game of "lupine roulette."

However, it does appear that some of the alkaloids found in lupines are removed by cooking and that toxins intensify with age. The toxic principle of lupines is excreted by the kidneys, and the poisoning is not cumulative. That is, a lethal dose must be eaten at one time to cause death. The poisonous effects produced by lupines are referred to as lupinosis, with nervousness, labored breathing, convulsions, frothing at the mouth the obvious signs (Muenschner 1940). But until documentation on Idaho species is established, lupines as a food source are not recommended. Many people use the larger, hairy leaved species as an excellent toilet paper substitute.

Alfalfa (*Medicago*)

General Description Three species occur in Idaho and include *M. lupulina* (black medic), *M. polymorpha* (burclover), and *M. sativa* (alfalfa). In general, they are described as hairless, branching perennial or annual herbs with leaves divided into three leaflets. The terminal leaflet is evidently longer than the other two. The pods are twisted. Usually found in disturbed areas at the lower elevations.

Field Notes/Uses Alfalfa can cause bloat in livestock when it constitutes a high percentage of their diet. Saponins found in the leaves may contribute to the problem (Kingsbury 1964). Humans should, therefore, use this plant in moderation. The dried and powdered young leaves and flower heads of alfalfa are nutritious (Peterson 1977), and can be steeped in hot water to make a bland tea. The tender leaves can also be added to salads and are rich in Vitamins A, D, and K. Alfalfa also supplies calcium, magnesium, and phosphorus (Kuhnlein and Turner 1991). Alfalfa sprouts are a popular salad addition and the seeds are available from various health stores. Nectar from the flowers produces a good honey. In addition to uses as food and medicine, Alfalfa seeds contain oil for use in paints and varnishes. Papermakers have used the stem fibers in their craft, and wool dyers extract a yellow dye from the seeds. The seeds of black medic can be parched and eaten, or ground into flour.

Sweetclover (*Melilotus*)

General Description Two species of sweetclover occur in Idaho and are strongly tap-rooted perennial or annual herbs. The leaves are divided into three fine-toothed wedge-shaped leaflets. The white or yellow flowers are loosely arranged in an inflorescence and the pods are thickly spindle-shaped. Usually found in disturbed habitats at the lower elevations. The genus name is from the Greek *mel*, meaning honey, and lotus flower.

Field Notes/Uses The young leaves (before the flowers appear) of and *M. officinalis* (yellow sweetclover) may be eaten raw or boiled. The fruit may be used as seasoning for soups (Weedon 1996). The older leaves are toxic and should be avoided. The dried flowering plant of *M. officinalis* was used in teas for neuralgic headaches, nervous stomach, diarrhea, and aching muscles (Peterson 1977). Elias and Dykeman (1982) indicate that improperly dried yellow sweetclover will easily mold and in the process produce coumarin, an anti-coagulant that can cause severe bleeding and death. Molding yellow sweetclover mixed in hay has killed many cattle. Poorly dried or fermented yellow sweetclover produces dicoumarol, a potent anticoagulant that is extremely poisonous in excess. It is used in rat poisons.

The plants are sweet scented due to coumarin and become more pleasant when dried. They have been used to scent clothes and protect them from moths as an alternative to moth balls. It has also been a traditional flavoring additive in smoking tobacco and snuff.

Stemless-loco (*Oxytropis*)

General Description Ten species of *Oxytropis* occur in Idaho. In general, they are perennial herbs that are stemless or with a short leafy stem. The flowers are white to reddish or purple. The keel is prolonged into a point or tooth or a straight to curved beak. The name is from the Greek *oxys*, meaning sharp, and *tropis*, meaning keel, in reference to the beaked keel. The species are found in a variety of habitats from the lower elevations to subalpine.

Field Notes/Uses These are the famous "locoweeds." The plants are poisonous to livestock. In order for the plants to be lethal to livestock, the animals must eat large amounts over a long period of time. Extensive grazing of *O. sericea* (white-flowered loco) induces a chronic poisoning called locoism. It is strongly recommended that these plants be avoided for the purposes of human consumption.

Goldenbanner (*Thermopsis*)

General Description Two species in Idaho: *T. gracilis* (slender goldenbanner) and *T. montana* (mountain goldenbanner). These are rhizomatous plants with trifoliate leaves. The stipules are often leaf-like and the petals are yellow in color. The yellow flowers somewhat resemble those of lupines.

Field Notes/Uses Ramah Navajo used a decoction of *T. montana* as cough medicine and created a fumigant of mountain goldenbanner for headaches (Vestal 1952).

Clover (*Trifolium*)

General Description Many species of clover can be found in Idaho. In general, they are annual and perennial plants from rhizomes with leaves that are

divided into three or more leaflets. The flower colors range from white, pink, yellow, red, or purple, and the seed pods are round to elongated. They are found in various habitats at all elevations. The genus name refers to the three leaflets.

Field Notes/Uses All species are nutritious and high in protein, but the flower heads and tender young leaves are hard to digest raw and may cause bloating. To improve digestibility of the plants, soak them in salt water for several hours or overnight. Leaves prepared this way may be dried and stored for future use. The dried flower heads and seeds can be ground into a flour substitute or extender (Peterson 1977).

Trifolium was an important food source for many Native Americans. In the spring, explorers and settlers saw them in the meadows picking and eating large quantities of clover. This was an annual event for the Natives, who relished the greens of the spring season. Unfortunately, the non-natives in their ignorance compared the Natives to grazing animals. This is just one of many disparaging comments made concerning misunderstood Native American behavior (Strike 1994).

A tonic tea can be made from the dried flowers. Made strong, the tea can be used as a gargle for sore mouths, throat, and a mild sedative (Willard 1992). The tea can also be used as a wash for skin ailments. The dried leaves can be smoked.

Vetch (*Vicia*)

General Description Five species of vetch occur in Idaho. These annual or perennial herbs have trailing to climbing stems. Leaves are pinnately divided with tendrils in place of terminal leaflets. Found in waste places at lower elevations. *Vicia* closely resembles *Lathyrus* (sweetpea) and requires close examination of the stipules. The stipules of *Vicia* are usually cut into narrow lobes, whereas the stipules of *Lathyrus* are entire to dentate.

Field Notes/Uses Many species contain toxic compounds and therefore should be considered poisonous, however, Kirk (1975) and Craighead et al. (1963) state that the young stems and seeds can be boiled or baked. The seeds of some species contain compounds producing toxic levels of cyanide when digested.

Caution Because of the poisonous compounds found in vetch, it is not recommended for eating.

Fumatory Family - FUMARIACEAE

There are 16 genera and 450 species within the Fumatory family. The family name is derived from *fumus*, the Latin word for smoke, and refers to the climbing, purple-flowered European species *Fumaria officinalis*, where the growth resembles a "cloud of smoke over the ground." This is due to the finely divided leaves. The members of this family are of little economic importance, except a few are cultivated as ornamentals. Members of this family are sometimes included in the Papaveraceae (Poppy family)

Corydalis (*Corydalis*)

General Description Two species of *Corydalis* occur in Idaho, *C. aurea* (scrambled eggs) and *C. caseana* (Sierran fumewort). They are annual or perennial herbs with dissected leaves. The flowers are yellow or white to pinkish in color. They can be found in moist, open areas and along streams.

Field Notes/Uses The plants are considered to be poisonous and contain several different alkaloids. Native American Indians apparently used a tea made from *C. aurea* for painful backaches, diarrhea, menstruation, bronchitis, sore throats, stomachaches, and inhaled the fumes of burning roots for headaches (Foster and Duke 1990).

Bleeding-heart (*Dicentra*)

General Description The two species of *Dicentra* in Idaho are *D. cucullaria* (dutchman's breeches) and *D. uniflora* (longhorn steer's-head). They are perennial herbs from tubers with dissected leaves, and white to purplish or rose-colored. They are usually found growing in well-drained soils from foothills to the subalpine.

Field Notes/Uses The plants are considered to be poisonous and contain several different alkaloids (Kingsbury 1964). These alkaloids are found throughout the plant and can cause trembling, staggering, convulsions, and labored breathing. Large quantities can be fatal. A poultice from *D. cucullaria* was apparently made to treat skin diseases (Foster and Duke 1990).

Drug Fumitory (*Fumaria officinalis*)

General Description This species is usually in disturbed places at the lower elevations. In fact, this annual is found practically everywhere on earth. The small flowers vary in color from reddish-purple to yellowish-white.

Field Notes/Uses The medicinal properties of the plant include its being a diuretic, laxative, tonic, and stomachic. It is used internally primarily for liver and gallbladder problems (Lust 1979). In large doses it works as a laxative and diuretic.

Gentian Family - GENTIANACEAE

There are 70 genera and approximately 1,100 species of this family worldwide. Thirteen genera are native to the United States. They are mostly annual or perennial herbs with bitter juice. Several species are cultivated as ornamentals.

Green Gentian (*Frasera*)

General Description Members of this genus are perennial herbs with opposite or whorled leaves. Flowers are bell-shaped and are densely aggregated into pyramid shaped panicles. The four species in Idaho can be found in dry, open areas or meadows up to the subalpine zone.

Field Notes/Uses The fleshy root of at least *F. speciosa* (showy frasera) can be eaten raw, roasted, or boiled. But, because the root is very bitter, we suggest mixing it with salad greens. An infusion of *F. albicaulis* (whitestem elkweed) was used to treat infected sores (Strike 1994).

Gentian (*Gentiana*)

General Description This is a large genus with most species occurring on moist or wet soil. Members are annual, biennial or perennial herbs from fleshy roots or rhizomes. The flowers are 4 or 5 lobed, tubular- or funnel-shaped. The five species can be found from the foothills to alpine meadows. The genus honors King Gentius of Illyria, ruler of an ancient country on the east side of the Adriatic Sea, who is reputed to have discovered medicinal virtues in gentians.

Field Notes/Uses Moore (1979) suggests that gentians are perhaps the best stomach tonics. As a bitter, gentians excite the flow of gastric juices, thereby promoting an appetite and aiding in digestion. The root or chopped herb is steeped and drunk before a meal. Herbage and roots of most species is bitter (Weedon 1996). Craighead et al. (1963), in discussing *G. calycosa* (Rainier pleated gentian) make mention of medicinal uses of European and Asian gentians, and that early settlers used them in much the same way as a tonic. Gentians contain some of the most bitter tasting compounds known, against which the bitterness of other substances is scientifically measured.

Dwarf Gentian (*Gentianella*)

General Description The three species in Idaho (*G. amarella*, *G. propinqua*, and *G. tenella*) are glabrous annuals with basal and cauline leaves. The calyx tube is shorter than the lobes, and the corolla has spreading lobes that are shorter than the tube.

Field Notes/Uses *G. propinqua* was used as a cold remedy. A decoction of the leaves, stems, and flowers were also taken for colds and cough. Another related species (*G. quinquefolia*) provided an infusion for treating diarrhea, and the liquid from the root was used for hemorrhages.

Geranium Family - GERANIACEAE

There are 11 genera and 800 species in this family distributed worldwide. *Geranium* and *Erodium* are native to the United States. Members of this family are 5-merous plants (5 petals, 5 or 10 stamens, pistil of 5 parts). The seedpod resembles head and beak of a stork or crane, hence the common name, with the seeds in the short thickened "head," and the style is elongated into the pointed "beak." The family is of no real economic importance, except as a source of ornamentals, primarily from the cultivated geranium (Pelargonium), a tropical genus well developed in South Africa.

Red-stem Storksbill (*Erodium cicutarium*)

General Description This is a low growing annual with mostly basal, finely dissected, fernlike, pinnately divided leaves. The flowers are small, pink and mature into the distinctive "stork's bill" fruit. This is an introduced plant that is widespread on disturbed sites at low to middle elevations.

Field Notes/Uses The leaves of red-stem storksbill can be eaten raw in salads or cooked as a potherb. They are particularly palatable when picked young, and have a parsley-like taste. We find it nicely compliments an otherwise

bland wild salad and provides a good source of Vitamin K. It is uncertain whether other species of *Erodium* are edible and it is not recommended.

The species has a reputation of being a diuretic, astringent, and anti-inflammatory herb. The entire plant was used in a warm water bath for persons suffering from the pains of rheumatism (Foster and Duke 1990). Leaves were also used in a hot tea to increase urine flow and to increase perspiration.

Wild Geranium (*Geranium*)

General Description The five species of geranium found in Idaho are annual or perennial herbs that are hairy. The leaves are mostly basal, and the flowers are showy, with 5 petals and sepals, and 10 stamens. The mature fruits are spirally coiled. They can be found in wet meadows or dry, open forests. *Geranium* is derived from the Greek word *gernion*, meaning "crane."

Field Notes/Uses The leaves and flowers of most species can be eaten, but because of their astringent properties and texture, they are not a choice edible. We find that they are best when tossed in with other greens in salads or steamed as potherbs. In any case, the leaves are better treated as a filler to stretch supplies of other more tasty and less abundant greens. The leaves can also be chopped and added to soups, thereby blending flavors making the leaves more acceptable. Leaves do

toughen with age, but are still palatable in stews. Geranium leaves are similar looking to monkshood (*Aconitum*), so positive identification of the flowerless plants is important. Harvest leaves and roots from plants identified with flowers.

The herbaceous part of *G. viscosissimum* (sticky geranium) was used as astringent and styptic, internally for diarrhea and hemorrhages (Willard 1992). The plant is high in tannins, providing astringent remedies important in traditional medicine for the emergency treatment of injuries and diarrhea. A hot poultice of boiled leaves was used for bruises and skin problems. The green crushed leaves can be applied to relieve pain and inflammation. A leaf or root tea of *G. richardsonii* (Richardson's geranium), which is one of the most widespread western species and frequently hybridizes with other species, can be used as a gargle for a sore throat. The root sliced fresh can be used as a first aid for gum or tooth infections when applied directly on the area of pain (Moore 1979).

Currant and Gooseberry Family - GROSSULARIACEAE

This family consists of a single genus (*Ribes*) with approximately 150 species. All are shrubs with palmately lobed leaves. Some species are armed with spines. The family is a source of ornamentals and edible fruits. In many old field guides, Ribes is sometimes included as a member of the Saxifragaceae (Saxifrage family).

Currant, Gooseberry (*Ribes*)

General Description The many species of *Ribes* in Idaho are shrubs. The species that have prickles on the stems and bristles on the fruit are

commonly called gooseberries. Those without prickles on the stem or bristles on the fruit are currants. Leaves are palmately veined and shallowly or deeply lobed. The five petals are smaller than the sepals and usually narrowed to a claw-like base. Fruit is a berry.

Field Notes/Uses The berries of almost all species of *Ribes* are edible raw, and none are known to be poisonous. However, we have come across some unpalatable species, berries with an unpleasant odor and a taste to match. The berries are high in Vitamin C and one of the richest plant sources for copper. One method of collecting them in bulk, is by shaking the bushes over sheets of plastic or blankets. Those that are too sour or spiny become more palatable if they cooked or dried. In regard to the fruits with bristles, one can also roll the berries on hot coals in a basket until the bristles have been singed off. When dried, the berries are a great trail snack. The dried berries can also be mixed with meat to make pemmican. The berries contain enough natural pectin to make jelly. The seeds also contain large quantities of gamma-linolenic acid and many herbalists use this oil to treat skin conditions, asthma, arthritis, and premenstrual syndrome (Willard 1992, Foster and Duke 1990). The nectar-filled flowers are considered good trail snacks (Clarke 1977). The wood makes good arrow shafts.

Leaves of currents and gooseberries may be added to herbal tea blends. The leaves should be fresh or thoroughly dried, not wilted as they may be toxic.

Water Milfoil Family - HALORAGACEAE

This family occurs throughout the world, but mostly in the Southern Hemisphere. In general, members of this family are aquatic herbs with simple or pinnatifid leaves that are opposite, alternate or whorled.

Water-milfoil (*Myriophyllum*)

General Description The genus name is from the Greek *myrios* (many) and *phyllon* (leaf), referring to the finely divided leaves. There are about 40 species of these submerged aquatic and terrestrial herbs with pinnate leaves and spikes of small, wind pollinated flowers. There are at least four species in Idaho.

Field Notes/Uses The rhizomes were frozen for future use, eaten raw, fried in grease, or roasted. The rhizomes are sweet and crunchy and a much relished food and were an important food during periods of low food supplies.

Mare's-tail Family - HIPPURIDACEAE

The family consists of a single genus, *Hippuris*. The genus was at one time assigned to the Haloragaceae (Water Milfoil family), but it is not closely related.

Mare's Tail (*Hippuris vulgaris*)

General Description This is a common plant in the main mountain chains of the western United States. The plant at first glance resembles an immature horsetail (*Equisetum*), but they are unrelated. Horsetails reproduce by spores and have stems that can be quickly pulled apart. The flowers of mare's tail are small and inconspicuous. The plant is found in the margins of shallow waters from ponds to lakes to streams. It can also be found in marshy and swampy areas, roadsides, and irrigation ditches.

Field Notes/Uses The whole plant is edible when prepared as a potherb. The plant parts are tender and can be gathered in any stage, even in winter. Ancient herbalists are said to have employed mare's tail for internal and external bleeding.

Hydrangea Family - HYDRANGEACEAE

There are approximately 17 genera and 250 species in this family mostly found in the Northern Hemisphere, from the Himalayas to North America. Nine of the genera are native to the United States. The family is a source of a few ornamentals.

Lewis' Mockorange (*Philadelphus lewisii*)

General Description This is the State Flower of Idaho and is also known by the common name of Syringa. Syringa is a shrub up to ten feet tall with leaves that are opposite and egg-shaped with three distinct veins. The flowers are white and when in full bloom, the flowers scent the air with a delightfully sweet fragrance reminiscent of orange blossoms. It occurs in the lower to middle elevations. The genus is named for the Egyptian King Ptolemy Philadelphus and the species name honors the scientist-explorer Meriwether Lewis, who first discovered and collected it during his expedition through the Louisiana Purchase.

Field Notes/Uses The wood of syringa is strong and hard, and does not crack or warp. It is an excellent wood for making bows and arrows. The leaves and flowers foam into lather when bruised and rubbed with hands, and can be used for cleaning the skin. The plant is otherwise considered poisonous.

Waterleaf Family - HYDROPHYLLACEAE

The 20 genera and 270 species within the Waterleaf Family are distributed worldwide, except for Australia. The western United States appears to be the main center of diversity. Only a few members in the family are cultivated.

Waterleaf (Hydrophyllum)

General Description Three species of waterleaf occur in Idaho. They are somewhat fleshy perennial herbs with leaves that are pinnately divided. Flowers are white to bluish. They are found in moist soils from the foothills to alpine environment.

Field Notes/Uses The young shoots, leaves, and flowers of *H. capitatum* (ballhead waterleaf), *H. fendleri* (Fendler's waterleaf), and *H. occidentale* (western waterleaf) can be eaten raw, or these and the roots may be cooked and eaten. We find them exceptionally good in salads, or when eaten as a trail nibble. They do have a texture that takes some getting used to.

The leaves can be used as a protective dressing for minor wounds, and are slightly astringent. As a poultice, it can be used for insect bites and other minor skin irritations.

Fiddleleaf (*Nama*)

General Description These are annual or perennial plants with leaves well distributed along the stems. The flowers occur in cymes and are axillary and the calyx is deeply divided.

Field Notes/Uses The seeds of a related species (*N. demissum*) were dried, pulverized, and boiled with water to make mush or porridge.

Baby Blue Eyes (*Nemophila*)

General Description Four species occur in Idaho. These are mostly annuals with stems that are diffuse, weak, and sometimes prostrate. Leaves are mainly opposite and the flowers occur in the upper axils of the leaves. The style is deeply bifid. The genus name is from the Greek for "grove loving," referring to the woodland habitat of many species in this genus.

Field Notes/Uses The roots were used to prepare a decoction to cure asthma by some Native tribes.

Phacelia (*Phacelia*)

General Description The many species in Idaho include herbaceous annuals, biennials, and perennials with various degrees of hairiness. Flowers are 5-parted, spirally coiled, with stamens extending beyond the corolla. The species occur in various habitats and elevations.

Field Notes/Uses At least one species, *P. ramosissima* (branching phacelia) can be cooked and used as greens (Weedon 1996, Kirk 1975, Craighead et al. 1963). However, Strike (1994), suggests that the stems and leaves of Phacelia may have been eaten raw, but were most likely cooked. The boiled roots of *P. ramosissima* were used to cure coughs and colds, and to alleviate lethargy. A decoction was also used as an emetic and to relieve stomach aches (Strike 1994). Medicinally, the whole plant of *P. heterophylla* (varileaf phacelia) was dried and pulverized, then used to poultice wounds, which facilitated healing (Strike 1994).

Mint Family - LAMIACEAE

There are approximately 180 genera and 3,500 species worldwide, with the Mediterranean region being the chief area of diversity. All have epidermal glands that exude an odor when rubbed, not necessarily pleasant and mint-like. Labiatae refers to corolla shape, 2 lips (labia). This family is of considerable economic importance as a source of numerous ornamentals, aromatic oils, and a few weedy genera as well as medicines.

Giant Hyssop (*Agastache*)

General Description The two species in Idaho are perennial herbs with flowers clustered in a dense spike-like inflorescence. The flowers are white to purplish. The species are common in meadows from foothills to subalpine.

Agastache is Greek (*agan*) meaning much, and *stachys* meaning "ear of grain," referring to the flower cluster.

Field Notes/Uses The seeds of *A. urticifolia* (nettleleaf giant hyssop) may be eaten raw or cooked, and the leaves can used as a tea or for flavoring stews. The Miwoks of California drank an infusion made from the leaves to relieve rheumatic pain, and for indigestion and stomach pains (Strike 1994, Train et al. 1957). The plant is said to have mild sedative qualities. Mashed leaves were made into a poultice for swellings (Train et al. 1957).

American Dragonhead (*Dracocephalum parviflorum*)

General Description This is a biennial or short-lived perennial from a taproot. The lower stem leaves are small and wither early, and the stalked upper leaves are lance-shaped with spine-tipped teeth. The purple flowers are small in a dense, spike-like inflorescence. Dragonhead occurs in open, low elevation moist sites. The plant requires disturbance, such as fire or scarification, to germinate. Another species, *D. thymni* (thymeleaf dragonhead) also occurs in Idaho.

Field Notes/Uses Kirk (1975) indicates that the Havasupai Indians of northern Arizona are reported to make flour from the seeds of American dragonhead.

Ground Ivy (*Glecoma hederacea*)

General Description This is a "rough to the touch" perennial. The leaves are heart-shaped to kidney-shaped, with rounded teeth. The flowers are violet-blue in the leaf axils. Though native to Eurasia, the plant has become established across the United States, and occurs in disturbed and moist woods at the lower elevations. Unlike other mints, ground ivy does not have any discernible odor.

Field Notes/Uses The plant is said to possess astringent and expectorant properties. Folk uses of the plant include a poultice used topically that is useful for bruises and muscle aches. It was also said to have been used in combating a variety of respiratory ailments, including bronchitis, pneumonia, and coughs.

Caution While there are no accounts of human poisoning, it appears to be toxic to horses that have consumed large amounts of the plant.

Henbit (*Lamium amplexicaule*)

General Description This is a weedy annual from Eurasia. The leaves are rounded to heart-shaped with shallow, rounded teeth on the margins. The flowers are pinkish-purple arising from the leaf-like bracts. It is found in waste places and fields at the lower elevations. *Lamium purpureum* (purple deadnettle) is also listed to occur in Idaho.

Field Notes/Uses The entire plant of *L. amplexicaule* is edible. We found it best when added to other salad plants. The plant is mildly astringent and a tea from the leaf and flower is used by herbalists for minor internal or external bleeding, to relieve diarrhea, and other digestive problems. A poultice from the fresh plant is said to reduce swelling, and help in the healing of minor burns, insect bites, and wounds.

Caution In most cases the plant may be found in areas that are often sprayed with herbicides and pesticides.

Common Motherwort (*Leonurus cardica*)

General Description Common motherwort is a fibrous-rooted perennial from a short rhizome. The leaves are palmately cleft and coarsely toothed. The flowers are pale pink in color and have an upper lip that is white-hairy. Motherwort is an Asian species that is sometimes encountered in fields at the lower elevations. The scientific name is from the Greek *leon*, meaning lion, and *oura*, meaning tail.

Field Notes/Uses Motherwort has a history as a Chinese and Western herbal medicine. It was used to tone the heart muscle and has been shown to slow and strengthen the heartbeat while lowering high blood pressure (Tilford 1997). It is also said to possess diuretic, sedative, and analgesic properties. Tilford (1997) indicates that the effectiveness and safety of motherwort is well documented in Chinese clinical and laboratory studies.

Bugleweed (*Lycopus*)

General Description The three species *Lycopus* in Idaho are perennial herbs with rhizomes. Of the three species, *L. americanus* (American waterhorehound) is the more common. The small flowers are pinkish-purple, whorled in the axils of the upper leaves. They occur in wet to moist areas of lakes and riverbanks. Unlike other mints, bugleweeds do not have the mint-like odor.

Field Notes/Uses The leaves of American waterhorehound are edible raw, but are usually tough and bitter. The tubers of *L. uniflorus* (northern bugleweed) can be added to salads, pickled, or boiled and eaten (Peterson 1977). Folk uses of *L. americanus* include its use as a cough remedy. It is also believed to have been used as a diuretic, mild sedative, and astringent in a tea or tincture form (Foster and Duke 1990).

Horehound (*Marrubium vulgare*)

General Description This is a woolly perennial herb with bitter sap. The leaves are wrinkled and toothed. The flowers are small, white and occur in dense whorls. Horehound is a common weed of waste places and fields at the lower elevations.

Field Notes/Uses Horehound is listed as a stimulant, tonic, expectorant, and diuretic (Moore 1979, Callegari and Durand 1977). The plant was highly valued by ancient Egyptian priests and the Romans, with the former calling it the "Seed of Horus" and "eye of the star."

The most famous use of this plant is horehound candy and is used to soothe sore throats and coughs. A tea from the dried leaves and flowers is also used, but because of the extreme bitterness of the herb, it is obvious why it tastes better in the form of a candy. Following is a recipe from Clarke (1977) that we have used on a number of occasions to make the horehound candy and cough syrup;

> "The plant can be boiled or dried without losing its flavor, which is unusual for a member of the mint family. One cup of fresh leaves or 1/4 cup dried leaves boiled in two cups of water for 10 minutes will make a strong concentrate. This concentrate can then be diluted with 2 parts water to 1 part concentrate for a tea. One part concentrate may also be added to 2 parts sugar or honey and a pinch of cream of tarter, brought to hard crach (290 degrees Fahrenheit), and poured into a buttered plate for the old fashioned cough drop candy. A bit of lemon added at the last minute improves the flavor. A cough syrup can be made of 1 part concentrate and 2 parts honey."

Other medicinal uses of the plant include making a warm infusion that will promote perspiration and the flow of urine. When taken cold, this infusion will expel worms.

Mint (Mentha)

General Description All three species in Idaho are distinctly aromatic perennial herbs with rhizomes. The flowers are arranged in whorls. *Mentha* comes from Mintho, the mistress of Pluto, ruler of Hades. Pluto's jealous queen, Proserpine, upon learning of Minthos, trampled her, transforming her into a lowly plant forever to be walked upon. Pluto made this horrible fate more tolerable by willing that the more the plant was trampled, the sweeter it would smell.

Field Notes/Uses The fresh or dried leaves of *M. arvensis* (wild mint), *M. canadensis* (Canadian mint), and *M. spicata* (spearmint) can be steeped in hot water for a tea. They have also been used as flavoring agents for soups, meat, and pemmican. The young leaves can also be added to salads and soups. The plants are high in Vitamin A, C, K, and minerals iron, calcium and manganese. It is an appetite stimulant and digestive aid. The leaf tea is considered medicinal and was used for colds, stomachaches, fevers, headaches, insomnia, and nervous tension (Foster and Duke 1990). Crushed leaves were used by some Native Americans to poultice swellings and bruises (Strike 1994).

Pure mint oil is a multimillion dollar industry. It is added to shampoos, massage oil, salves, and soaps, as well as medicines, foods, and liqueurs. It takes approximately 300 pounds of mint to yield 1 pound of oil.

Herb Teas

Plants are the source of practically all the beverages consumed by mankind. Water and milk are the exception. They provide flavor, color, and aroma for endless variety and pleasure. They also provide nutrients for health.

Most herbal teas are infusions made by pouring boiling water over herb leaves or flowers and allowing them to steep for 5-10 minutes to release the herb's aromatic oils. A general rule is one teaspoon of dried herb, or 3 teaspoons of fresh crushed herb, per cup of water. To make a stronger tea, add more of the herb rather than steeping the tea longer (long steeping makes the tea bitter). Experiment by combining various herb teas for interesting flavor results.

Beebalm (Monarda fistulosa)

General Description The plant is from a spreading rhizome, and the leaves are sharp-tipped and glandular. The flowers are lavender or rose-purple and aggregated in dense terminal heads. The sepals are fused into a ribbed tube, the top of which has five short, spiny lobes covered with dense, white hairs. It is found on open, dry to moist slopes and on rockslides and forest openings at the lower elevations.

Field Notes/Uses The young leaves and leaf buds of beebalm can be used as a seasoning much like you would use oregano, in salads. Moderation, however, is advised. While the older plants can also be used, they are very bitter and it would be best to dry them first.

Pacific Monardella (*Monardella odoratissima*)

General Description This is an aromatic perennial that has a woody base. Flowers are in dense head subtended by broadly ovate bracts. It is found on dry slopes at moderate elevations. *Monardella glauca* (pale monardella) also occurs in Idaho and could be used in much the same way as Pacific monardella.

Field Notes/Uses The leaves and stalks of Pacific monardella were eaten, and a thirst quenching tea was made from the leaves and flower heads (Strike 1994). Medicinally, the tea made from the inflorescence of *M. odoratissima* was used for colds and fevers, and relieved digestive upsets and purified the blood (Strike 1994, Weedon 1996). The tea made from the whole plant used in the first stages of a cold when fever is present is said to help relieve elevated temperatures and toxins through sweat (Moore 1979). Leaves and flowers rubbed on exposed skin repels mosquitoes and other biting insects.

Catnip (*Nepeta cataria*)

General Description This is a tap-rooted perennial that feels like felt to the touch. Leaves are triangular-shaped and coarsely toothed. The flowers are blue or yellowish-white in terminal, spike-like inflorescences. This introduced species from Europe is now widespread across North America and occurs in waste and disturbed places, and along irrigation canals at the lower elevations.

Field Notes/Uses The nutritious young leaves and buds can be added to salads. The dried leaves and flowers make an excellent tea and are high in trace minerals and vitamins. The tea is a subtle, relaxing sedative on humans. As with other mint teas, it is also a carminative, and is soothing to an upset stomach.

Common Selfheal (*Prunella vulgaris*)

General Description Common selfheal is a cosmopolitan plant of moist areas, from meadows to fields up to the subalpine. The flowers are whorled in a dense head-like cluster. It is also called Hercules' All-heal, because it is supposed that Hercules learned the herb and its virtues from Chiron.

Field Notes/Uses The entire plant is edible, raw or cooked. However, we found that it is the young and tender plants collected in the early spring that are best. The crushed leaves can be used fresh or dried to make a tea.

Historically, as the common name implies, the plant has been used as a medicine for almost everything. Herbal uses include an astringent, antispasmodic, tonic and styptic (Willard 1992). The tea of the dried plant was also used as a gargle for sore throat (Foster and Duke 1990). Fresh plants can be made into an antiseptic poultice for bruises and scrapes (Tilford 1993), because of the high tannin content.

Sage (*Salvia*)

General Description The three species of *Salvia* in Idaho are *S. dorrii* (grayball sage), *S. pratensis* (introduced sage), and *S. sclarea* (Europe sage). Members of this genus can be annual or perennial herbs or shrubs. Look for sage at the lower elevations.

Field Notes/Uses The seeds of perhaps all species of Salvia may be eaten raw, or parched and ground into meal. The seeds can also be soaked in water for a flavorful drink. Leaves of any fragrant sage can be used as a tea or spice for soups and meats. Sage does contain moderate amounts of Vitamin A and C, and can be added fresh to salads and sandwiches, however, we advise you to do this sparingly.

Tilford (1993) indicates that the above ground parts are antiseptic, astringent, hemostatic, alterative, and tonic, and make a good strong topical disinfectant and cleansing wash for abrasions, contusions, and chafed skin (Moore 1979). They are also an effective gargle for sore throat and congested sinuses (Moore 1979).

Note *Salvia* and some species *Artemisia* are often mistaken for being a "sage." Both are aromatic, but are plants in different families. Sage (*Salvia*) is a member of the Mint family, while wormwood (*Artemisia vulgaris*) and sagebrush (*Artemisia tridentata*) are in the Sunflower family. Mints have opposite leaves, whereas these Artemisia species have alternate leaves.

Skullcap (*Scutellaria*)

General Description Four species of *Scutellaria* occur in Idaho. In general, they are perennial herbs with axillary flowers and are found in wet or moist soils in the middle elevations.

Field Notes/Uses *Scutellaria galericulata* (marsh skullcap) has nervine related therapeutic properties and has been used as a remedy for general restlessness (Willard 1992). It has been used in acute or chronic cases of nervous tension or anxiety. The calming effects are said to be mild but reliable (Tilford 1997). A strong tea was also made of *S. lateriflora* (blue skullcap) and used as a sedative, nerve tonic, and antispasmodic for all types of nervous conditions. All the species contain scutellarin, the primary active compound that has been confirmed to be a sedative and has antispasmodic qualities (Foster and Duke 1990). Other species may have similar qualities.

Hedge-nettle (*Stachys palustris*)

General Description This is a hairy, glandular perennial from a rhizome. Hedge-nettle has no stinging hairs, as do the true nettles (*Urtica*), but resembles them before flowering. The flowers are white-dotted, reddish-purple in several whorls. It is common on wet sites at lower elevations. If cultivated, it should be from seed.

Field Notes/Uses The leaves and flowers are edible, but because of their fuzzy texture and bitter taste, we find them unpleasant. The tubers can also be eaten raw, cooked or pickled and are best if collected in the autumn. Other species may be edible, but this has not been confirmed. Strike (1994) indicates that *Stachys* tubers were eaten by Native Americans.

The leaves may be soaked in water for a few minutes and used as a poultice (Kirk 1975). Additionally, an infusion of fresh leaves can be used as a wash for sores and wounds.

Mountain Blue Curls (*Trichostema oblongum*)

General Description This is an annual plant with soft hairs. The flowers are violet, and deeply lobed. It can be found growing on moist soils at the lower elevations. The genus name arises from the Greek *trichos* (hair) and *stemon* (stamen), referring to the long slender stamens, a characteristic of the genus.

Field Notes/Uses While there are no recorded uses for Mountain Blue Curls, a related species, *T. lanceolatum* (vinegarweed), has a strong pungent vinegar odor and was best known as a fish poison. The fresh plants would be mashed and thrown into the pools or sluggish streams. The intoxicated fish would then float to the surface where they were easily caught by hand. Strike (1994) indicates that the fine hairs on the flowers would catch on the gills of the fish and interfere with respiration, making the fish easier to catch. Medicinally, the leaves were chewed and put in the cavity of an aching tooth (Sweet 1976). Moore (1979) says that a tea made from the flower tops is good for stomachaches and promotes sweating in dry fevers.

Bladderwort Family - LENTIBULARIACEAE

There are five genera and 300 species in this family worldwide. Two genera, *Utricularia* and *Pinguicula* are native to the United States, and occur in Idaho. They are described as annual or perennial herbs of moist and aquatic habitats. The insectivorous species in this family, trapping by means of sticky leaves and bladders, are sometimes cultivated as oddities.

Bladderwort (*Utricularia*)

General Description More than 250 species in the United States and four occur in Idaho. These are aquatic or bog plants with submersed stems. The leaves are finely dissected and the yellow flowers are strongly two-lipped with a spur at the base. They are found growing in ponds, lakes, and sluggish streams at the low to middle elevations.

Field Notes/Uses Bladderworts are carnivorous plants that entrap small aquatic animals in their bladders. The bladders are closed at the narrow end by valve-like doors that have stiff trigger hairs on the outer surface. When set, the bladders have a partial vacuum, and when a passing animal touches the bristles, the doors open, the walls of the bladder immediately expand, and the sudden inrush of water captures the prey. The process has been timed at 1/460 of a second. Enzymes then digest the trapped victim.

The edibility and uses of bladderworts is unknown, but several species in this genus are reputed to have diuretic values and have been used to treat dysentery (Strike 1994, Coon 1974).

False Mermaid Family - LIMNANTHACEAE

The two genera and 12 species in this family are restricted to North America. They are annual herbs found in wet places.

False Mermaidweed (*Floerkea proserpinacoides*)

General Description This is a slender annual with succulent stems. The leaves are pinnately divided into 3-5 oblong leaflets. The white flowers are stalked and borne singly in the leaf axils. The plant is found in moist, shaded habitats, especially under shrubs. The genus is named after a German botanist,

H.G. Floerke. This plant is so inconspicuous that only those who are familiar with it are apt to notice it.

Field Notes/Uses We had sampled the stems and leaves of this plant and found them spicy. They were an acceptable addition to our wild salad.

Flax Family - LINACEAE

There are about 12 genera and 300 species in this family worldwide. The family is of some economic importance because of flax fibers, linseed oil, and the ornamentals obtained.

Flax (*Linum*)

General Description The three species in Idaho are either annual or perennial herbs with yellow or blue flowers. Flax has had value through the ages for its many uses, such as for thread, fabric, oil, paper money, and cigarette paper.

Field Notes/Uses The seeds contain a cyanide compound, but are edible after roasting them (Kirk 1975). They have a high oil content that contains essential fatty acids that are very much needed in our daily lives (Willard 1992), plus they add an agreeable flavor to cooked foods. The crushed seeds have been used as a poultice for irritation, boils, and pain (Sweet 1976). An infusion of stems is said to relieve stomach aches or intestinal disorders (Strike 1994). The roots were also steeped to make an eye medicine. The stems are a source of linen, a fabric used for clothing.

Northwest Yellow Flax (*Sclerolinon digynum*)

General Description This annual plant with yellow flowers and oblong to elliptic and mainly opposite leaves occurs in moist, grassy meadows.

Field Notes/Uses Seeds are probably edible after cooking.

Mentzelia Family - LOASACEAE

There are about 15 genera and 250 species of this family occurring chiefly in South America and the warmer parts of the North America. Various species of *Mentzelia* are endemic in the western United States.

Blazingstar (*Mentzelia*)

General Description Seven species of blazingstar occur in Idaho. There are many species of *Mentzelia* in the western United States. It is also called "stick-leaf" because of the barbed hairs on the leaves which readily cling to fabric.

Field Notes/Uses *Mentzelia* was considered an important food source in many places of the West. The seeds are edible after being parched and ground into flour (Olsen 1990, Kirk 1975, Bean and Saubel 1972). Murphy (1990) describes a type of "gravy" made from the seeds of *M. albicaulis* (whitestem blazingstar) and *M. laevicaulis* (smoothstem blazingstar);

> "...the red seed is put into a hot frying pan and when the seeds turn a darker red, warm water is added and it is stirred till it thickens."

The Hopi Indians in the southwest parched and ground the small, oily seeds of *M. albicaulis* into a fine, sweet meal and ate it in pinches.

Loosetrife Family - LYTHRACEAE

Members of this family are herbs, shrubs, or trees. The leaves are opposite or whorled. There are approximately 25 genera and 550 species widely distributed around the world. Seven genera are native to the United States. The family is a source of dyes and ornamentals.

Loosetrife (*Lythrum*)

General Description Two species occur in Idaho: *L. salicaria* (purple loosetrife) and *L. tribacteatum* (threebract loosetrife). Purple loosetrife is a Eurasian species that has become a nasty wetland weed. It is also cultivated in gardens by some who are unaware of its potential. It is often called the "beautiful killer" because it can take over wetlands and displace native species. It appears to have some efficacy against gnats and flies, and was reported to calm quarrelsome beasts of burden at the plow if placed upon the yoke (Pojar and MacKinnon 1994).

Field Notes/Uses A tea made from whole flowering plant of purple loosetrife, fresh or dried, is a European folk remedy for diarrhea, dysentery, and a gargle for sore throats. It was also used as a cleansing wash for wounds. Experiments have shown that the plant extracts stop bleeding, and kill some bacteria (Foster and Duke 1990). Other species appear to have been used by Native Americans. For example, *L. californicum* was used by the Kawaiisu Indians in California as a medicine and as a dermatologic aid. The method, however, is not reported. Additionally, *L. hyssopifolia* was used by Native Americans in California to expedite healing and to reduce inflammation of mucous membranes. It was also used as a shampoo for the hair, but the method is not reported.

Mallow Family - MALVACEAE

There are some 85 genera and 1,500 species in this family, most of which occur in the tropics. Twenty seven genera are native to the United States. The distinctive feature of this family is the uniting of the numerous stamen stalks to form a tube around the pistil that resembles a tree trunk, with the anthers and non-fused filaments as the branches and leaves. This "stamen tree" in the center of the flowers is almost a "never fail" characteristic of this family. The family is of moderate economic importance because of cotton fibers derived from the seeds of *Gossypium*, several ornamentals, and a few food plants. Mallow is from the Greek word meaning soft and may refer to the soft fuzzy leaves characteristic of so many plants in this family, or to the sticky, soothing juice obtained from the roots of some species.

Streambank Globemallow (*Iliamna rivularis*)

General Description This is a perennial herb up to four feet tall. All green parts of the plant have star-shaped hairs. The flowers are lavender to pink in color and in dense clusters. The plant is usually found along stream courses, moist sites, and on disturbed sites such as clearcuts.

Field Notes/Uses The stems of some species were chewed as gum (Craighead et al. 1963).

Mallow (*Malva*)

General Description Four species of mallow occur in Idaho. The plants are distinguished by their distinctive fruit and seeds, rather than their leaves

and flowers. They are introduced annual or biennial herbs that are usually found in waste places at the lower elevations.

Field Notes/Uses The entire plant of *M. neglecta* (dwarf mallow) is edible. The young leaves are particularly good in salads or cooked up as a potherb. The plant is, however, very mucilaginous, and it is often used to thicken soup and may take a little getting used to. Eaten in large amounts, however, may cause digestive disorder. The immature fruits (which look like cheese) can also be eaten raw or added to soups.

Medicinally, the bruised leaves of *M. parviflora* (cheeseweed mallow) can be rubbed on the skin to treat skin irritations. As a headache remedy, leaves or the whole plant can be mashed and placed on the forehead (Willard 1992). Leaf or root tea can be used for angina, coughs, bronchitis, and stomachaches (Foster and Duke 1990). The fresh or dried leaves were used as a soothing poultice (Moore 1979).

Checkermallow (*Sidalcea*)

General Description There are two species in Idaho. Checkermallows are annual herbs with lobed or divided leaves. Flowers are white to deep pinkish-lavender, in terminal clusters. Both species can be found in meadow-type habitats.

Field Notes/Uses Kirk (1975) says *S. neomexicana* (New Mexico checkermallow) is edible after cooking as greens.

Globe Mallow, False Mallow (*Sphaeralcea*)

General Description Three species of globe mallow can be found in Idaho. They are perennial herbs that have star-shaped hairs on the leaves and stems. Flower colors range from red to pink. They can be found in open areas at the lower elevations. The genus name comes from the Greek *sphaira* and *alkea* to mean "spherical mallow."

Field Notes/Uses *Sphaeralcea coccinea* (ccarlet globemallow) was chewed and applied to inflamed sores and wounds as a cooling, healing salve. It was also used as a pharmaceutical aid. The entire plant was ground and steeped in water for a sweet tasting tea that was mixed with other bad tasting medicines to make them more palatable. The Navaho Indians used the plant as a lotion to treat skin diseases, as a tonic to improve appetite, and as a medicine for rabies.

Buckbean Family - MENYANTHACEAE

Members of this family are perennials with thick rhizomes, usually found in aquatic habitats. The leaves are simple or divided into three sessile leaflets. There are five genera and 30-40 species distributed worldwide.

Buckbean (*Menyanthes trifoliata*)

General Description The one species in Idaho, *M. trifoliata*, is a perennial marsh herb with creeping root-stocks. The leaves have long petioles and are all basal and divided into 3 leaflets. The whitish flowers are small, star-shaped, and crowded into a short inflorescence. The species can be found in bogs and lakes. The species is sometimes placed in the Gentianaceae (Gentian family)

Field Notes/Uses The herbage and rhizome of the plant are bitter, but we found that the rhizome can be made palatable when collected in early season and boiled in several changes of water. Nutritious flour can also be made from the rhizome by drying, crushing, and leaching it thoroughly. The fresh plant eaten raw may cause vomiting.

The dried leaves are tonic, diuretic and are esteemed, due to high content of Vitamin C, iron, and iodine (Foster and Duke 1990). Buckbean tea was used to relive fever and migraine headaches, for indigestion, to promote a

healthy appetite, and to eliminate intestinal worms (Pojar and MacKinnon 1994). A poultice of leaves can be applied to skin sores, herpes, glandular swelling, and for sore muscles (Willard 1992). Fresh leaves are an emetic, therefore dry them well before use unless you intend to induce vomiting. The plant contains a bitter glycoside, menyanthine, which stimulates gastric juices. The leaves have been used in facial steams for those troubled with achene. Add the tea to a bath or use as a rinse for oily hair. The fruit has no known use. Leaves are a common ingredient in herbal smoking blends.

Carpetweed Family - MOLLUGINACEAE

These are often succulent and perennial herbs with whorled or opposite leaves. Petals are usually absent in the flowers, and the fruit is a many seeded capsule. The family is chiefly distributed in South Africa with approximately 130 genera and 2500 species. Other centers of distribution include Australia, Asia, South America, and the western United States. The family is of little economic importance, except that a few species are eaten as potherbs.

Green Carpetweed (*Mollugo verticillata*)

General Description This one species is an annual with whorled leaves. The flowers are whitish and without petals. Carpetweed is a weed of fields and waste places found at the lower elevations.

Field Notes/Uses Only one reference for the uses of this plant could be found. Kirk (1975) indicates that the plant may be used as a potherb.

Waterlily Family - NYMPHAECEAE

There are six genera and 68 species found throughout the world in aquatic habitats. Four of the genera are native to the United States, and are a source of food for birds and aquatic animals. The family name (Nymphaeceae) translates as water nymph or water virgin. It was once believed that the plants in this family had anti-aphrodisiacal properties and were then used in art to represent virginity. A few species are used in cultivation.

Yellow Pondlily (*Nuphar lutea*)

General Description This plant is easier to identify than to harvest. The plant has mostly floating leaves. The flowers are of obvious bright yellow sepals, and the petals are inconspicuous and yellowish-green to purplish-tinged. It is common in shallow portions of many ponds and lakes of low to middle elevations.

Field Notes/Uses The rhizomes are best during the early spring and fall. These starchy rootstalks can be boiled and then peeled and eaten or placed in soup or stew, or dried, ground into meal and used as flour. The plant reproduces by seeds and rhizomes and is very easy to culture.

The seeds can be collected and, when dry, will keep indefinitely. They can also be treated like popcorn. Simply pop them and eat or grind them into meal. The seeds can also be steamed as a dinner vegetable or cooked like oatmeal -- 1 part seeds, to 2 parts water. In Turkey, the flowers of another species are distilled into a beverage called pufer cicegi.

The leaves and stalks have been used as poultices for boils, ulcerous skin conditions and swelling. An infusion of the root is useful as a gargle for mouth and throat sores. Some aboriginal people still use a root medicine for numerous illnesses, including colds, internal pains, rheumatism, chest pains, and heart conditions (Pojar and

MacKinnon 1994). According to Spellenberg (1979), pondlilies produce alcohol:

> "...when the mud in which the stems grow loses oxygen, a small amount of alcohol instead of carbon dioxide is produced."

Evening Primrose Family - ONAGRACEAE

There are about 20 genera and 650 species worldwide, with a dozen genera native to the United States. The family is of little economic importance, but a few are considered to be ornamentals. However, oil of evening primrose is obtained from this family and is said to be the world's richest source of natural unsaturated fatty acids. The oil is helpful in cases of obesity, mental illness, heart disease, and arthritis and is advertised widely in natural food publications.

Small Enchanter's Nightshade (*Circaea alpina*)

General Description Enchanter's nightshade is a delicate perennial growing up to 12 inches tall, with glandular hairs in the upper portions and very slender rootstocks. Heart-shaped leaves are 1 to 3inches long, with subentire to sharply toothed margins and narrow, pointed tips. The flowers are inconspicuous with tiny, 2-petaled corollas. It is a circumboreal species and it is most often associated with cool, moist and shaded sites. Despite the name, it is not an alpine species; rather it occurs predominantly in montane and lower subalpine
Field Notes/Uses A related species, *C. lutetiana* (broadleaf enchanter's nightshade), was used by the Iroquois as a dermatological aid on wounds. They also made an infusion as a wash on injured parts (Chamberlain 1901).

Clarkia (*Clarkia*)

General Description Two species of *Clarkia* occurin Idaho: pinkfairies (*C. pulchella*) and diamond clarkia (*C. rhomboidea*). They are annuals with brittle stems and purple or red, showy flowers. They are usually found on dry slopes at the lower to middle elevations. The genus honors Captain William Clark of the Lewis and Clark Expeditions to the Northwest in 1806.
Field Notes/Uses Sweet (1976) indicates that the seeds of *C. pulchella* were dried, parched, pulverized, and then eaten by some Native Americans. Additionally, Strike (1994) states that *Clarkia* seeds were prized by many California Natives and were often used to make pinole. The roots of many *Clarkia* species were also eaten.

Willowherb, Fireweed (*Epilobium*)

General Description Many species of willowherb occur in Idaho found in various habitats and elevations. The genus includes annual and perennial plants that have willow-like leaves. The flowers are white or lavender in color with petals that are often notched. Fruits are long, narrow pods that open by 4 slits to release the numerous small, densely hairy seeds. The roots and pods are often needed to make positive identification of the many species. The genus name is from the Greek, meaning "on a pod," describing the elongated ovary bearing the other flower parts on its top. The common name refers to the tufts of hairs at the end of the seed, which is similar to that on willow seeds.

Field Notes/Uses The dozens of species of *Epilobium* are all reported to be edible for people caught in survival situations, but *E. angustifolium* (fireweed) and *E. latifolium* (dwarf fireweed) are the best known and most commonly utilized species. Recent molecular studies have resulted of both these species being moved into the genus *Chamerion* (*C. angustifolium* and *C. latifolium*, respectively).

Food, drink, tinder, twine, and medicine are all provided by these abundant herbs. There are many small and "weedy" species found. In general, they are survivors in landscapes that have been ravaged by man-made and natural forces (e.g., fires, clearcuts). Soil conditions do appear to affect their flavor. Many Native Americans "owned" good patches of fireweed and these were passed on to subsequent generations. The most distinctive identifying feature of fireweed is the unique leaf venation. Unlike other plants, the veins do not terminate at the edges of the leaves, but rather join together in loops inside the outer margins.

The young shoots and leaves of fireweed may be boiled like asparagus, but are better when mixed with other raw greens for a salad. The leaves, green or dry, make a good tea and are useful in settling an upset stomach (Tilford 1993). Be careful, the leaves are slightly laxative. The unopened flower buds can be used in the same manner as leaves and stems. The young fruits can also be boiled like green beans and are tasty before the seed fibers form. Mature plants tend to become tough and bitter.

The pith of fireweed stems can also be scraped out and eaten as a snack or as a thickener for soups. If consumed in large amounts, fireweed is a gentle but effective laxative. The plant contains a relatively high content of Vitamin C and beta-carotene. Raw roots are a popular food of Siberian Eskimos. A poultice made from the roots of fireweed can be used on skin inflammations, boils, ulcers, and rashes (Moore 1979).

The fibrous inner bark can be used as cordage and tinder material. For use in making cordage, I found the fibers brittle. The seeds have cotton-like hairs and are great for fire starting (tinder) and insulation. Many Northwest Indians used the fluffy seed cotton as a wool substitute, mixing it with mountain goat wool or duck feathers. Willow-herb fluff, however, lacks the qualities of a really fine fiber. The flowers can also be rubbed into rawhide to repel water.

Groundsmoke (*Gayophytum*)

General Description There are six species of *Gayophytum* in Idaho. They are slender-stemmed annuals with alternate leaves, the lower ones often being opposite. The flowers are small and white. The various species are found on dry slopes, and on the edges of meadows.

Field Notes/Uses An infusion of *G. ramosissimum* (pinyon groundsmoke) was used to soothe irritated skin (Strike 1994).

Evening Primrose (*Oenothera*)

General Description Many species in this genus occur in Idaho. They are annual, biennial, and perennial herbs. The flowers are white or yellow, often opening at night. There are 8 stamens, 4 petals, 4 sepals, and the stigma is

globe-shaped to deeply four-lobed. The various species can be found in a variety of habitats up to the subalpine zone.

Field Notes/Uses Most handbooks on edible plants indicate that at least *O. elata* (Hooker's evening primrose) and *O. biennis* (common evening primrose), which have edible roots. These were cooked and eaten as a vegetable when young, becoming tough and somewhat spicy or peppery with age. The leaves of *O. biennis* are also edible as cooked greens, but are not exceptional unless mixed with bland greens to make a more acceptable salad. Harrington (1967) suggests that all species would stand a trial as none are known to be poisonous. The various species are known to hybridize easily making identification at times challenging.

We have cooked and eaten the young seed pods of several species and found them to have an acceptable taste. Olsen (1990) also suggests that many species have seeds that are edible after being parched or ground into meal. Strike (1994) states that seeds and leaves of Oenothera were eaten by Natives Americans.

The leaves, stems, and crushed seeds have an astringent quality to them and can be used as a poultice to heal wounds, and for bruises and piles. The seeds are also high in essential oils and have been shown in clinical studies to be effective for heart disease, asthma, arthritis, alcoholism, and other fatty acid problems (Willard 1992, Moore 1979). The medicinal uses of the oil in these plants are a recent discovery following scientific research in the 1980's that demonstrated their effectiveness for a wide range of intractable complaints. The oil contains gamma-linoleic acid (GLA), an unsaturated fatty acid, which assists in the production of hormone-like substances. Evening primrose oil, in the form of gelcaps, is becoming popular in the natural supplements marketplace. Additionally, the stringy bark makes good cordage material.

Broom-rape Family - OROBANCHACEAE

There are 13 genera and 180 species found around the world, with four of genera being native to the United States. Members of the Broom-rape family are herbaceous, lack chlorophyll, and are parasitic on the roots of other flowering plants. The family is of no direct economic importance. The family name comes from the Greek orogos, meaning "a clinging plant" and acho, "to strangle." Recent molecular studies have greatly expanded this family to include members of the figwort family usch as paintbrush (Castilleja).

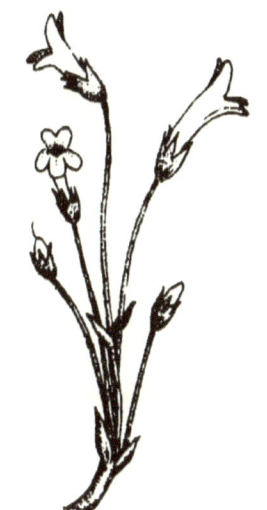

Broom-rape (Orobanche)

General Description All six species in Idaho parasitize the roots of other plants. These fleshy annual plants are nearly white to brownish or purplish in color and lack chlorophyll. The leaves are reduced to scales. Broom-rape is usually found in dry soils, associated with such genera as Artemisia and Eriogonum.

Field Notes/Uses The entire plant of broom-rape, roots and all, can be eaten raw. Being succulent plants, they answer for food and drink, and are often called "sand food." We found them to be better tasting when roasted in the hot ashes of a campfire. Strike (1994) also indicates that the roots of *O. californica* (California broom-rape) and the entire plant of *O. fasciculata* (clustered broom-rape) were eaten.

The decocted blanched or powdered seeds are said to ease joint and hip pain. They can also be used as tooth ache remedy (Willard 1992). Moore (1979) indicates that the whole plant is astringent and would make an excellent poultice. Broom-rape is also mildly laxative and has sedative properties. The stalks with the white inner portions removed have been used as pipes. *Orobanche uniflora* (one-flowered broom-rape) was used to treat numerous ailments including bronchial problems, intestinal upset, toothaches, and rheumatic pain (Strike 1994). A decoction of *O. fasciculata* was used as a skin wash to kill lice.

Wood-sorrel Family - OXALIDACEAE

There are seven genera and over 1,000 species distributed worldwide. Only *Oxalis* is native in the United States. In general, they are small plants with leaf blades divided into 3 heart-shaped segments. Flowers are 5-merous, and yellow or purple. The seedpods split explosively, scattering seeds some distance from the plant. The family is of little economic importance.

Woodsorrel (*Oxalis*)

General Description The three species in Idaho are described as above. The genus name is derived from the Greek word oxys, meaning "sour".

Field Notes/Uses The leaves and stems of *O. corniculata* (creeping woodsorrel), *O. stricta* (common yellow oxalis) and *O. trilliifolia* (three-leaf woodsorrel) may be eaten raw. To make a tasty dessert, collect and allow a mass of the plants to ferment for a while. The plants also contain a high percent of oxalic acid; therefore, it is recommended that one eat the plants sparingly until accustomed to them. One symptom of too much oxalate is painful or swollen taste buds. The plants are also high in Vitamin C and were used to remedy scurvy. A drink can be made by steeping the leaves in hot water, followed by chilling, and sweetening it (Peterson 1977).

Oxalic Acid

The tart, lemony taste of wood sorrel *(Oxalis)*, cacti *(Opuntia)*, lamb's quarters *(Chenopodium)*, amaranth *(Amaranthus)*, knotweed *(Polygonum)*, dock *(Rumex)* and other species, is due to the presence of soluble oxalic acid. Without proper preparation these plants when eaten in substantial amounts should be considered toxic. However, when properly prepared, these plants are an excellent food source.

The soluble oxalic acid, also known as salt of lemon, is what makes the plants tasty as well as dangerous. The oxalic acid is dangerous because of its solubility and its affinity for calcium. The solubility allows the acid to enter the bloodstream, where it promptly combines with calcium to form non-soluble calcium oxalate. This precipitates in the kidneys where it both plugs the tubules and "burns" all cells in contact with it, potentially leading to renal failure and death.

The oxalic acid is readily dissolved in heated water and will combine with calcium as readily in that water as in the bloodstream. Adding bone fragments, egg shells, or some other sources of calcium to cooking water will transform the oxalic acid to non-soluble calcium oxalate in the pot, retaining the full flavor, but rendering harmless the acid.

If you have no bone fragments or egg shells, just pour out the first water after boiling for a time and replace with fresh water.

Peony Family - PEONIACEAE

The family has only one genus with approximately 33 species. Brown's peony is a perennial herb with thick roots. The leaves are ternately divided or compound. The flowers are large and showy.

Brown's Peony (*Paeonia brownii*)

General Description This is a perennial herb with thick roots. The leaves are ternately divided or compound. The flowers are large and showy. The genus name is from the Greek for Paeon, the physician of the gods who supposedly used the plant medicinally.

Field Notes/Uses Northwestern Indians made a tea from the roots to treat lung ailments (Strike 1994, Spellenberg 1979). Train et al. (1957) also indicate that Peony was used as an internal treatment. A decoction of boiled roots was taken for tuberculosis, venereal disease, coughs, and nausea. The root decoction was also used as a lotion for headaches, a liniment for swellings, a gargle for a sore throat, and a wash for sore eyes. The mashed root was used as a poultice for boils and deep cuts or wounds. The pulverized dried root was used for cuts, wounds, sores, and burns.

Poppy Family - PAPAERVERACEAE

Most plants in this family are annual or perennial herbs, and sometimes shrubs. The sap is often milky or colored. There are 26 genera and 200 species in this family distributed in the subtropical and temperate areas of the northern hemisphere, particularly in western North America. Thirteen genera are native to the United States. The family is of little economic importance except for *Papaver somniferum* which yields opium and its many derivatives, including morphine and heroin. A few species are cultivated as ornamentals.

Prickly Poppy (*Argemone*)

General Description Two species occur in Idaho, *A. munita* (flatbud prickly poppy) and *A. polyanthemos* (crested prickly poppy). These are annual or perennial herbs armed with prickles that are usually found growing in dry places at the lower elevations. The sap may be yellow, orange, or milky.

Field Notes/Uses The seeds of *Argemone* have been used in the past as food, but they are so difficult to extract that it hardly seems worth it. Medicinally, the ripe seeds of *A. munita* were roasted, mashed, and applied as a salve on burns and abrasions by some Native Americans. The seeds were also pounded and used as a poultice on open sores and as a hemorrhoid remedy. The juice of Argemone has a rubifacient and somewhat caustic effect and was used for burning off warts (Moore 1979). A tea made from the plant is an analgesic topically and can be applied to sunburns and abrasions to relieve pain and swelling.

Warning The plants contain toxic alkaloids.

California Poppy (*Eschscholzia californica*)

General Description This is an annual plant with ternately dissected leaves. The flowers are orange or yellow and the two sepals are shed just after flowering. The plants in Idaho have escaped cultivation and are usually seen at the lower elevations.

Field Notes/Uses The flowers and leaves were eaten by some Native Americans (Strike 1994). The foliage was eaten by gathering the plants before the plants bloomed, leaching them in running water, and then cooking them.
Sap from the fresh root is mildly narcotic. It was apparently used by the Cahuilla Indians in California as a sedative for babies (Bean and Saubel 1972). Additionally, a piece of the root was placed in a tooth cavity to stop toothache. A root extract was used as a wash or liniment for headaches and open sores (Strike 1994). If taken internally, the root extract caused vomiting.

Warning All parts of the plant should be considered toxic.

Plantain Family - PLANTAGINACEAE

Three genera and 270 species of this family are found worldwide. Plantago is widespread in the United States. In general, the family is of little economic importance, but several species of *Plantago* are weeds and one (*P. psyllium*) is the source of seeds used to make a commercial laxative. Recent genetic studies have greatly expanded the size of this family. Members of the figwort (Scrophuraliaceae) and others are now found here.

Plantain (*Plantago*)

General Description Six species occur in Idaho and are characterized as short-stemmed annual or perennial herbs with basal leaves. The flowers are greenish or purplish. Many of the species are introduced from Europe and can be found at the lower elevations, particularly in fields and waste places. *Plantago* means "sole of foot" and refers to the sole shaped leaves of plantain that lie close to the ground as though stepped on.

Field Notes/Uses Because of their reputation as being weeds, plantains are a forgotten edible to many people. In fact, a lot of effort is spent trying to get rid of the plants from gardens. *P. major* (common plantain) and *P. lanceolata* (narrowleaf plantain) were brought over by European settlers for use as potherbs and medicine (Gibbons 1964). The Native Americans called the plants "white man's foot" because they followed the settlers west. The native species of plantain are uncommon in comparison and it is suggested that they only be used when large populations are found.

As a food, the young leaves of common plantain and narrowleaf plantain were used fresh or cooked. They contain calcium and other minerals. One hundred grams of plantain is said to furnish as much Vitamin A as a large carrot (Tull 1987). The older leaves may be too fibrous and bitter for use, but they are usable if one is able to remove the fibers. Seeds are tedious to collect in quantity, but can be ground and used as flour substitute or extender.

The leaves and seeds of many species were used medicinally. The foliage contains tannins and iridoid glycosides, notably aucubin, which stimulates uric acid secretion from the kidneys. The crushed leaves of common plantain provide an astringent juice that can be used to soothe wounds, sores, insect bites, and the rash of poison ivy (Tull 1987). The plantain juice is a traditional treatment for earaches. The seeds contain up to 30% mucilage which swells in the gut to act as a bulk laxative and soothes irritated membranes. Rubbing the leaves on the skin works as a natural, moderately effective insect repellant.

Phlox Family - POLEMONIACEAE

There are about 18 genera and 320 species in this family, found chiefly in North America and particularly in western United States. The family is a source of a few ornamentals.

Trumpet (*Collomia*)

General Description These are annual or perennial herbs with simple or branched stems. There are approximately 15 species in western North America and South America. The flowers are funnel-shaped or tubular with throats that abruptly flare into an expanded limb. The genus name is from the Greek, kolla, meaning glue, because of the mucilaginous layer on the seeds of most species.

Field Notes/Uses The seed coat becomes mucilaginous when wet. This "glue" helps keep the germinated seeds from drying out. This is a mechanism that helps store water between the first autumn rains and those that may not come for several weeks.

From the roots of large-flowered collomia an infusion was made for high fevers. Additionally, an infusion of the leaves and stalks was taken for constipation and to "clean out the system." Narrow-leaved collomia was used as a dermatological aid by the Gosiute. They made a poultice of the mashed plant and applied it to wounds and bruises.

Gilia (*Gilia*)

General Description There are many species of gilia in Idaho and are characterized as annual, biennial, or perennial plants. The leaves are mostly alternate, lobed or dissected with the tips acute. The seeds are sticky when wet. They are found in a variety of habitats at various elevations.

Field Notes/Uses Strike (1994) indicates that Gilia seeds were eaten by many Native Americans.

Scarlet Gilia, Skunk Flower (*Ipomopsis aggregata*)

General Description Seven species of Ipomopsis occur in Idaho, but only *I. aggregata* (skyrocket gilia) has recorded uses (see illustration). Skyrocket gilia is a biennial plant 1-3 feet tall. The tubular or funnel-form flowers are red, orange, pink, or white and showy. The plant is usually found on dry slopes up to the subalpine. The species has also been called skunk flower because of a faint skunk-like smell from its glandular foliage.

Field Notes/Uses The plant was used by Native Americans as a tea to treat colds, to make glue, and to treat blood troubles. In Nevada, the principal use of this plant was for the treatment of venereal diseases. The whole plant was boiled for the purpose and a solution was taken as a tea or used as a wash (Coffey 1993). The whole plant was also boiled by the Ute Indians in Utah to make glue (Densmore 1990). A blue dye can be extracted from the roots.

Granite Gilia (*Leptodactylon*)

General Description Three species occur in Idaho. They are low shrubs with prickly leaves. The flowers are funnel-form to salver-form and often very showy. The species are found in dry, rocky places in the lower to high elevations.

Field Notes/Uses *Leptodactylon pungens* (granite prickly gilia) was used as a decoction to bathe swellings, sore eyes, and scorpion stings (Strike 1994).

Gilia (*Linanthus*)

General Description Three species on Linanthus occur in Idaho. In general, they are low annuals with opposite leaves that are palmately parted into slender segments or reduced to linear blades. They are found on dry, open slopes.

Field Notes/Uses While there are no recorded uses for Idaho species, an infusion was made from *L. ciliatus* by the Maidu and Pomo Indians in California to treat children's coughs and colds. An unheated decoction was drunk instead of water to purify blood (Strike 1994).

Slender Phlox (*Microsteris gracilis*)

General Description Slender phlox is an annual with linear to elliptical-linear leaves that are sessile. The lower leaves are opposite and the upper are alternate. The flowers are white, pink or lavender. This is a very common and very variable species. It is found in a variety of mostly open areas at the low to middle elevations.

Field Notes/Uses Slender phlox was eaten by Miwoks in California as greens. Maidu also used slender phlox as a poultice on bruises and wounds (Strike 1994).

Pincushionplant (*Navarretia*)

General Description These are annual plants with rigid stems. The sessile flowers occur in spiny densely bracted heads, and the calyx is cleft to base. The stigma is 2- or 3-lobed.

Field Notes/Uses The seeds of a related species, skunkweed (*N. squarrosa*), were gathered and dried in the sun, and then stored. When needed, the seeds were parched and then eaten dry. Medicinally, skunkweed was used as a tonic, fever reducer, laxative, and dye. The seeds of *N. leucocephala* (which occurs in Idaho) were also eaten, and a decoction of the plant was used to reduce swellings.

Phlox (*Phlox*)

General Description The plants in this genus are low shrubs, perennials, or annuals with opposite leaves. The flowers are salver-form in shape. The many species can be found in various habitats at all elevations.

Field Notes/Uses A decoction from the roots of *P. longifolia* (longleaf phlox) and other species were used by some California Native Americans as an eyewash for sore eyes (Strike 1994). The scraped roots were soaked in water or steeped or boiled to make the wash (Train et al. 1957).

Jacob's Ladder (*Polemonium*)

General Description These are perennials from woody rootstalks. The flowers occur in terminal or axillary cymes. One species has small, white flowers, while the other 3 are blue-flowered. All are malodorous to various degrees.

Field Notes/Uses The genus name is thought to honor Polemon, an early Greek philosopher. The common name of Jacob's ladder is based on the arrangement of the leaflets, and comes from the story in Genesis 28:12 of Jacob's dream of a ladder connecting heaven to earth. Medicinally, a decoction of Jacob's-ladder (*P. pulcherrimum*) was used as a wash for the head and hair.

Buckwheat Family - POLYGONACEAE

In the Buckwheat family there are 40 genera and 800 species, of which 15 genera are native to the United States. The family is best represented in the western states. The economic products include food plants and a few ornamentals.

Wild Buckwheat (*Eriogonum*)

General Description These are annual or perennial herbs; some species are woody at the base. The flowers are small and usually bright colored. The many species of *Eriogonum* can be found in various habitats at all elevations. The genus name is from the Greek erion, meaning wool, and gony, meaning knee or joint, referring to the hairy stems of many species.

Field Notes/Uses None of the species are known to be poisonous. The flowering stems can be eaten raw or cooked before they have flowered. Seeds can be collected (though tedious) and ground into flour. A tea from the root of Eriogonum was used to treat headaches and stomach problems (Strike 1994).

The plants are mildly astringent and were used as a gargle for sore throats (Moore 1979).

Alpine Mountain Sorrel (*Oxyria digyna*)

General Description This is a low perennial with simple roundish leaves clustered at the base of the stem. The flowers are small, red or greenish. The plant is found in cold, wet places among rock crevices at the higher elevations. The plant resembles miniature rhubarb, with small rounded leaves. It has always been highly esteemed in Arctic regions as a "scurvy-grass" with an agreeable sour taste.

Field Notes/Uses Perhaps one of the most refreshing plants one encounters in the high country is the alpine sorrel. The new growth up to flowering time can be eaten raw, when it tastes like mild rhubarb. The stems and leaves can be used in salads or prepared as a potherb. Some aboriginal peoples have been known to ferment mountain sorrel as a kind of sauerkraut. This is accomplished by simply letting the plant(s) sit in water for a while. This sauerkraut can then be stored for winter use. The plants were also dried in the sun for traveling. Plants are high in Vitamin C and can be used to prevent and cure scurvy. Large amounts could, however, cause oxalate poisoning (Tull 1987).

Knotweed (*Polygonum*)

General Description The many species of knotweed in Idaho are annual or perennial herbs with stems that are more or less swollen at the nodes. The flower colors include white, greenish, or pink. They can be found in various habitats up to the higher elevations.

Field Notes/Uses Experimentation may be the rule for *Polygonum* as none of the species are known to be poisonous. They do, however, vary in degrees of palatability. Tannins are found in the plants and large amounts might cause digestive upset and possible kidney damage (Coffey 1993, Willard 1992, Kirk 1975). In moderate quantities, however, the genus is generally regarded as safe. Based on our experiments with various species, some have peppery tasting leaves that can be used in flavoring foods. Others have starchy roots that may be eaten raw or boiled, and roasted. Still others have young foliage made into good salads or potherbs. In our opinion, of all the species, *Polygonum [Bistorta] bistortoides* (bistort) tastes the best. This species is very common in mountain meadows.

The seeds have been used whole or ground into flour. The seeds of Polygonum are described as a prehistoric food source and are frequently found in archeological remains (Kindscher 1987).

A decoction of the roots can be made for a sore mouth or gums. The root can also be used as an astringent, diuretic, antiseptic and alterative (Willard 1992). The roots were eaten by maritime explorers to prevent scurvy. There is a traditional European "Easter pudding" made up of bistort, nettle, and dock, all of which are high in Vitamin C.

Dock (Rumex)

General Description The nine species of dock in Idaho are annual or perennial herbs. They have small flowers that are greenish and aggregated in a large terminal inflorescence. The fruits are called utricles. They can be found in many habitats in the mountains.

Field Notes/Uses The young leaves of dock can be used as greens and we have found that the flavor varies from species to species. The young leaves are best when collected before the flower stalk emerges. Also, because the leaves become watery when cooked, use very little water and don't overcook them. The older leaves may be too bitter for use. Euell Gibbons (1964) found that the leaves of dock are high in Vitamin C and contain more Vitamin A than carrots. Native Americans ground dock seeds and used the meal to make breads. However, removing the papery seed cover involves a lot of work, and depending on the species, is probably more work than it is worth. The  distinctive sour taste of these plants is due to oxalic acid. As with other species that contain oxalic acid, docks should be used in small portions as they can cause calcium deficiency.

Poisoning from *Rumex* has only been recorded in livestock after large quantities were eaten. Medicinally, the crushed leaves can be applied to boils and the juice of leaves used to treat ringworms and other skin parasites. The juice of the plant and a poultice of the leaves have also been applied to the rash and pain caused by stinging nettles. A poultice of leaves was used for nervous or allergic hives. The fresh roots were boiled in water to provide a decoction for use internally as a laxative. The powdered yellow roots have been used as a tooth cleanser, a laxative, astringent, and an antiseptic (Lewis and Elvin-Lewis 1977). Some Rumex roots contain as much as 35% tannin and were used for tanning animal hides.

Purslane Family - PORTULACACEAE

Nineteen genera and 600 species occur worldwide, of which nine genera are native to the United States. The family is particularly well represented along the Pacific Coast. It is of little economic importance, but includes several ornamentals.

Fringed Redmaids (*Calandrinia ciliata*)

General Description This is an annual with linear to oblanceolate leaves, and red flowers. The plant is found in open, grassy areas at the lower elevations. The genus is named for J.L. Calandrini, born in Switzerland in 1703.

Field Notes/Uses The seeds of *C. ciliata* were prized by a great many Native Americans. After gathering and winnowing, the seeds were parched with coals, pulverized, and then pressed into balls and cakes for eating (Strike 1994). The roots were also eaten, and the young leaves and stems were eaten raw or cooked.

Pussypaws (*Cistanthe*)

General Description The leaves of these plants are alternate or basal, and spatulate in shape. The small flowers have sepals that are membranous or membranous-margined, and there are 2 or 4 petals. In older references, the species has been included in other genera such as *Calyptridium* and *Spraguea*.

Field Notes/Uses The stems of pussy-paws thermoregulate; that is, respond to temperature. When it is cool (evening and morning), the stems lay flat on the ground, protected from the chilling winds. As the day warms, the flower stems rise, absorbing the midday warmth, but moving away from the scorching earth. The tiny black seeds are an important food for rodents.

Spring Beauty (*Claytonia*)

General Description Many species occur in Idaho and are described as perennial succulent herbs arising from deep corms or fleshy taproots. The leaves are opposite and the flowers white or pink in color. The genus is named for Dr. John Clayton, a botanist and notable plant collector of Colonial days.

Field Notes/Uses Often called Indian Potato, Wild Potato, or Mountain Potato, the small corms can be eaten raw, boiled, or roasted. For many Native Americans, spring beauty was an important "root vegetable." When collecting, keep only the largest corms and replant the others. At first, many find the corms distasteful, as they do take a little getting used to. The corms are high in starch and, when cooked, taste like potatoes. Boil or bake the corm for 30 minutes. Most species are not plentiful, so be conservative in your endeavor. They can also be dried on strings for long-term storage.

The rosettes can also be eaten raw or cooked and are high in Vitamins A and C. They are better when mixed with other salad plants. The leaves of *C. sibirica* (Siberian springbeauty) were soaked and applied to the head as a remedy for a headache (Pojar and MacKinnon 1994).

Claytonia perfoliata (miner's lettuce) is one of the few native plants of the United States that has been cultivated elsewhere. Introduced into Europe, miner's lettuce is now cultivated and used for salads and as a potherb.

Bitterroot (*Lewisia*)

General Description Six species are known to occur in Idaho. Generally, they are small perennials from thick fleshy roots. There are 4-18 petal-like sepals, white to deep red in color. They occur in various habitats including gravelly, dry soils, in the vicinity of late lasting snows, and mountain meadows up to the upper subalpine zone.

Field Notes/Uses Although all species may be edible, L. rediviva (bitterroot) is the species that has been used extensively. These plants were an important food item for many Natives. The root is remarkably large and thick for a small plant, and contains nutritious farinaceous matter that is much prized. The roots are dug up in spring before flowering. Once dug, the root is peeled promptly and the small red "heart" (embryo of next years growth) is removed to reduce the roots bitter flavor. It is then steamed, boiled, or pit-cooked and eaten. The root can also be dried and will keep for a long time. The bitterness of the root varies and cooking is said to improve the flavor. The root boiled to a jellylike consistency will be pink in color. The pounded root was chewed for a sore throat (Sweet 1976).

Though some still collect it today, bitterroot is considered a rare plant in many areas. There is little evidence that harvesting by Native Americans has contributed to the rare status. Overgrazing and trampling by range livestock and habitat destruction from agricultural encroachment seem to have been a major impact on *Lewisia* populations. Remember, digging the roots destroys the plant. Programs to maintain and enhance habitat for the plant are recommended.

Sacajawea's bitterroot (Lewisia sacajaweana *B.L. Wilson & E. Rey-Vizgirdas*) *is the first plant species to be named in honor of Sacajawea. This is an Idaho native and occurs nowhere else in the world but in the south-central Idaho mountains. Just over two dozen populations of Sacajawea's bitterroot are known to exist, roughly three-fourths of them on the Boise National Forest. Scattered populations also occur on the Payette, Sawtooth, and Salmon-Challis National Forests.*

Sacajawea's bitterroot was not always considered a unique species. It was originally known as Kellogg's bitterroot, which is also found in California's Sierra Nevada Mountains. Recent work regarding the plant's genetic and physical characteristics confirmed that the Idaho plants are indeed distinct from the Sierra Nevada plants. Sacajawea's bitterroot is smaller than its California relatives are, and about half the size of the common bitterroot.

Miner's Lettuce (*Montia*)

General Description The genus is comprised of slightly succulent annual and perennial herbs. The flowers have two persistent sepals and five white or pinkish petals. Most Montia species grow in moist or seasonally wet areas that are partially to fully shaded habitats.

Field Notes/Uses All species of *Montia* have stems and leaves that can be eaten raw or boiled like spinach. The roots are also edible raw or boiled. In California, some Native Americans picked miner's lettuce and placed it near the nests of red ants. The ants were allowed to crawl over the leaves and were then shaken off. The residue left on the leaves by the ants had an acerbic flavor (Strike 1994).

A tea from the leaves was used as a laxative. A poultice made from the plant was used for rheumatic pains and to stimulate a poor appetite (Strike 1994).

Purslane (*Portulaca oleracea*)

General Description This is a small succulent annual herb found at the lower elevations. It has been used as a food for more than 2,000 years in India and Persia. In Europe, it is grown as a garden vegetable. The genus name may be derived from portula meaning "little gate," referring to the lid on the capsule.

Field Notes/Uses The stems and leaves of purslane have a tart taste. The entire aboveground part of the plant can be boiled, steamed, fried or pickled. The mucilaginous juice of the stems makes a good thickener for soups. Since the plant tends to hold a lot of dirt and grit, you may want to wash it thoroughly. Besides the good flavor, purslane also provides vitamins A and C, iron and calcium. The tiny black seeds are also nutritious. They can be ground and mixed with other flours.

Primrose Family - PRIMULACEAE

Worldwide, there are approximately 28 genera and 800 species found in this family. Eleven of the genera are native to the United States, mostly to the eastern part of the country. They are of minor economic importance as a source of ornamentals.

Scarlet Pimpernel (*Anagallis arvensis*)

General Description This is a small annual with opposite leaves that clasp the stem and are oval shaped. The small scarlet colored flowers are on solitary stalks. This is an introduced plant from Europe, and in some places of the United States it is quite common and weedy. Look for it in waste places at the lower elevations. The genus name is from the Greek Anagelao, which means to laugh. Dioscorides, physician to the Roman army, was said to give the plant to the men to relieve the depression that accompanies disorders of the liver. Another common name applied to this plant is "Poor Man's Weather Glass" because of its habit of closing before a rain.

Field Notes/Uses Scarlet pimpernel is said to be an effective diuretic which helps eliminate gravel from the kidney and is used in dyspepsia. As a poultice, it was applied to the skin to relieve the itch and sting of insects.

Rockjasmine (*Androsace*)

General Description The genus name is from the Greek *andros* (make) and *sakus* (a buckle), alluding to the shape of the anther. The genus contains about 100 species of alpine, annual and perennial herbs with simple, often tuft forming leaves and clusters of showy flowers. The petals are fused at the base. The genus differs from *Primula* in that the petal tube is shorter than the sepals and somewhat constricted at the mouth.

Field Notes/Uses A compound decoction was made from the whole plant of western androsace (*A. occidentalis*) and used for post-partum bleeding. A cold infusion of northern androsace (*A. septentrionalis*) was taken for internal pain. The plant was also used as a lotion to give protection from witches.

Shooting Star (*Dodecatheon*)

General Description The four species in Idaho are perennial herbs. All leaves are basal and form a loose rosette. The flowers are located at the end of a stalk with narrow, reflexed rose colored petals. The species habitats range from grassland to shrubland, meadows, and riparian habitats up to the alpine zone.

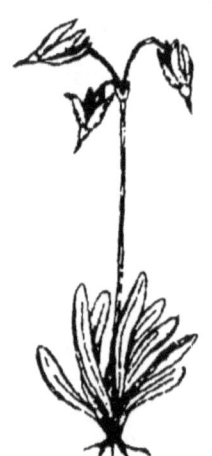

Field Notes/Uses Since none of the species are listed anywhere as poisonous, it is likely that all the species are edible. It is usually the texture that discourages people from using the plants. We have found that the leaves of many species have a good flavor when eaten raw. Weedon (1996) and Strike (1994) also indicate that the roots and leaves of *D. hendersonii* (Mosquito Bills) are edible after roasting or boiling. Scully (1970) believes that at least five species of shooting star in the Rocky Mountains were used by American Indians and that they ate the green leaves and roasted the roots. Thompson and Thompson (1972) provide some additional insight into their preparation of *D. jeffreyi* (tall mountain shooting star);

"...we tried eating the leaves of Sierra shooting star raw, but decided that their texture made them unappealing to chew on. At least they do not seem to become bitter, even after the flowers are blooming. When boiling for about 15 minutes and seasoned with butter and salt, they make a satisfactory but bland green vegetable."

Very little information regarding the medicinal uses of shooting star could be found. The Native Americans of the Northwest used a leaf tea as a treatment for cold sores (Tilford 1997).

Sea Milkwort (*Glaux maritima*)

General Description This is a perennial herb with rhizomes. The leaves are entire, sessile, and somewhat succulent. It is usually found in wet or moist alkaline or saline meadows, seeps, along distches and streams up to about 7,00 feet. This plant differs from all other genera of the Primulaceae in having apetalous flowers with a pink, petaloid calyx.

Field Notes/Uses The rhizome is edible after prolonged boiling and the young leaves can be used as a flavoring agent. The plant was traditionally consumed with grease, and only before bedtime, due to drowsiness effects.

Warning May cause sleepiness and/or nausea if eaten in quantity.

Tufted Loosestrife (*Lysimachia thyrsiflora*)

General Description This is one of three species found in Idaho. It is an erect perennial with opposite leaves that are sessile and gland-dotted. The yellow flowers occur in spike-like racemes and are deeply 5-parted into linear segments. The plant is occasionally found in wet places. The flowers are unique enough to make this plant pretty easy to identify. All of the Lysimachia species were formerly in the Primulaceae, but have been moved to the Myrsinaceae (Myrsine) family.

Field Notes/Uses The genus name is from the Greek lysis (releasing) and mache (strife). King Lysimachos of Thrace (c. 360-281 BC) was said to have pacified a bull with loosestrife. Some species are grown as ornamentals, others have local medicinal uses. For example, the Iroquois made a compound decoction of the plant and used it as a wash and applied it as a poultice to stop milk flow.

Primrose (*Primula*)

General Description This is a large group of plants commonly known as the Primrose. They are mostly hardy perennial herbs and are natives of Europe and temperate Asia, Java and North America. The generic name of the primrose is Primula from the diminutive of the Latin word *Primus*, meaning first.

Field Notes/Uses The young leaves of some primroses were used in salads. The leaves of these related species contain many vitamins. Additionally, some species are said to possess curative properties that the ancient Greeks once used in the treatment of coughs, tuberculosis, rheumatism, and insomnia.

A number of different insects visit the flowers in search of nectar, which is located at the bottom of the flower tube. However, this positioning means that only long-tongued insects can actually reach the nectar in the base. Next time you come across a primrose, take a close look at flowers and you may notice something very interesting. In fact, you will examine something that Charles Darwin once studied. It appears that primroses have two types of flowers: long-pistil flowers where the pistil is approximately twice as long as the stamens, and; short-pistil, where the stamens are approximately twice as long as the pistil. This feature is a means by which the plants are assured of cross pollination. In nature, both flowers types can be found with equal frequently. While cross pollination of two flowers of same type (and even a self-pollination) is possible, it is the cross pollination of flowers of the different types that produce more viable seeds.

Starflower (*Trientalis*)

General Description Two species occur in Idaho: *T. borealis* (broadleaf starflower) and *T. europaea* (arctic starflower). They are low, perennial herbs with tubers, and leaves crowded near the apex. The flowers are white or pinkish.

Field Notes/Uses The genus name is Latin, meaning "having one-third of a foot," refering to the plants' height. The sap of *T. borealis* was mixed with water and used as eyewash (Strike 1994).

Buttercup Family - RANUNCULACEAE

Worldwide, there are 35-70 genera and 2,000 species in the cooler regions of the Northern Hemisphere. Twenty one genera are native to the United States. Many plants in this family are poisonous, some are grown as ornamentals, and others provide drugs.

Note Only a few plants in this family were eaten. Most contain an irritating compound, protoanemonin in the fresh leaves, stems, roots, flowers, and seeds. All must be cooked before eating.

Columbian Monkshood (*Aconitum columbianum*)

General Description It is a perennial herb with palmately divided or lobed leaves. Flowers are usually deep blue or purple, but may also be pale to white. Usually found in moist, densely shaded places often with streamside vegetation up to timberline.

Field Notes/Uses Some species of monkshood have been a source of drugs, as a pain killer or as a sedative for nervous disorders. However, all parts of the plant are poisonous and should be considered dangerous if ingested. The drug aconite from *A. columbianum* was used to treat pain from neuralgia, toothache, and sciatia. Acotine is one of the most toxic plant compounds known. Nevertheless, a number of different species are used medicinally in various parts of the world, with apparently beneficial therapeutic effects.

Red Baneberry (*Actaea rubra*)

General Description It is a perennial herb with fibrous roots. The leaves have long petioles and are 2-3 times divided into sharply toothed, lance-shaped segments. The small flowers are white and borne in a branched, congested, hemispheric inflorescence. The fruits are shiny red or white. Red baneberry is common in moist, montane forests and riparian areas, usually with some partial shade.

Field Notes/Uses The entire plant, especially the berries, is poisonous. The plant is sometimes confused with *Osmorhiza chilensis* (western sweetroot) which often shares the same habitat. However, unlike red baneberry, sweetroot has a strong licorice-like odor.

The roots are considered a laxative and can cause vomiting (Peterson 1977). The roots were also ground, mixed with grease or tobacco and rubbed on the body to treat rheumatism. Ground seeds mixed with pine pitch were applied as a poultice for neuralgia (Coffey 1993). *Actaea arguta* is described by Moore (1979) as moderately poisonous when taken internally, with cardiac arrest possible from large doses. The powdered root was mixed with hot water and applied as a counterirritant.

Warning If large quantities of this plant are consumed, it may cause a cardiac arrest.

Windflower, Anenome, Pasque-flower (*Anemone*)

General Description The seven species in Idaho are described as perennial herbs from a rootstock or rhizome. The basal leaves are palmately lobed or divided, while the stem leaves are in whorls, each with 2-4 compound or simple leaves. There are no petals in the flowers, but the five to many sepals resemble petals. The plants can be found in dry to moist meadow areas from the foothill to alpine zone.

Field Notes/Uses Native Americans made a poultice from the leaves of A. cylindrica (candle windflower) to treat rheumatism and burns (Coffey 1993). The roots of A. globosa (=A. multifida) were used for treating wounds (Craighead et al. 1963). *Anemone patens* contain a volatile oil used in medicine as an irritant (Craighead et al. 1963).

Columbine (*Aquilegia*)

General Description Three species of columbine occur in Idaho. They are perennial herbs with large divided leaves. The lemon-yellow, reddish, blue or white flowers are showy with petal-like sepals. Each of the 5 petals extends

backward between the sepals forming a hollow spur. Columbines are found along mountain streams and open areas up to the alpine zone. The common name is from the Latin *columbinus*, and refers to the flowers resemblance to a cluster of doves. The genus name stems from the Latin name for eagle. It may also stem from the Latin *aqua*, water, and *legere*, to collect, perhaps referring to the nectar that collects at the tips of the spurs.

Field Notes/Uses The flowers of A. *formosa* (western columbine) are edible and have a sweet taste. They can be added to salads in small amounts. Weedon (1996) indicates that the leaves of this species are also edible, but grow bitter with age.

A tea made from the roots of western columbine is said to stop diarrhea and the fresh roots can be mashed and rubbed on aching joints (Sweet 1976). Aboriginal peoples used various parts of the plants in medicinal preparations for diarrhea, dizziness, aching joints, and possibly venereal disease (Pojar and MacKinnon 1994). The root, boiled with *Ipomopsis aggregata* (scarlet gilia) resulted in a brew that induced vomiting. Ripe seeds can be mashed and rubbed into the hair to discourage lice.

Warning The seeds can be fatal if eaten and most parts of the columbine contain cyanogenic glycosides. Any therapeutic use of columbine is strongly discouraged.

White Marsh Marigold (*Caltha leptosepala*)

General Description The leafless flowering stem of this plant has one or two flowers with 5-15 white or blue-tinged, petal-like sepals. There are no petals. The leaves are dark green and basal. It can be found in marshes and wet meadows to above timberline. Marsh marigolds bloom close to receding snowbanks. Marsh marigold is not related to the cultivated Marigold a member

of the Sunflower Family. Its nickname dates back to the Middle Ages when a plant (*Caltha*) was dedicated to the Virgin Mary and widely used in church celebrations. The flowers were fermented for making wine. The genus name is from the Greek calathos, meaning a cup.

Field Notes/Uses The young leaves of

C. leptosepala can be used as a potherb and the spaghetti-like roots can be dug up during the winter and boiled as a pasta substitute. Though the plant is poisonous when raw (the plant contains the poisonous glucoside, protoanemonin), cooking appears to destroy the poison. It also contains the deadly glucoside hellebrin, which breaks down with boiling (Mitchell and Dean 1982).

The roots have diaphoretic, emetic, and expectorant properties (Willard 1992). The leaves are diuretic and laxative, and a tea from the leaves mixed with maple sugar was used as a cough syrup by the Ojibwas (Willard 1992). The tea was also used as an antispasmodic and expectorant for treating cramps, and convulsions. In 'Stalking the Helpful Herbs', Euell Gibbons reports a drop of juice squeezed from the fresh leaves is caustic and will remove warts.

Bur Buttercup (*Ceratocephala testiculata*)

General Description Formerly known as *Ranunculus testiculatus*. This is a very tiny but very noticeable invasive species. It may appear in small patches but often densely carpets large areas. It flowers early in the spring when only a half inch tall; the flower gives way to a ¼ inch layered horny seed pod that dries and becomes very prickly. Bur buttercup has become widespread in gardens, pastures, waste areas, and along roadsides.

Field Notes/Uses This plant contains ranunculin, which changes into a highly toxic compound, protoanemonin, when the plant is crushed. Sheep have been poisoned and have died in the western United States after ingesting aboveground plant material. This plant is considered highly toxic. Signs of poisoning include anorexia, labored breathing, diarrhea, dyspnea, recumbency, weakness, and death.

Clematis (*Clematis*)

General Description Five species within this genus occur in Idaho. They are herbaceous perennials with erect stems or woody vines. The leaves are opposite or whorled and simple to pinnately compound. The flowers, lacking petals, are solitary or borne in an open, pyramid-shaped inflorescence. Sepals are petal-like. The various species can be found from brushy slopes above creek bottoms to open areas from the low to high elevations.

Field Notes/Uses The genus is essentially comprised of poisonous species. Many references list *C. ligusticifolia* (western white clematis) as poisonous even though the stems and leaves have been chewed by Native

Americans as a remedy for colds and sore throats (Tilford 1993). The plants have a peppery taste and may cause lightheadedness. Tilford (1993) also indicates that western white clematis is diaphorhetic, diuretic, and offers unique vasocontrictory/dilating action that makes it useful in the treatment of migraine headaches. The Thompson Indians used the plant to make a head wash for scabs and eczema, and a mild decoction is drunk as a tonic (Teit 1930). Sweet (1976) indicates that the white portion of the bark was used for fever, the leaves and bark for shampoo, and a decoction of the leaves was used on horses for treating sores and cuts. The fibers in the bark was used for snares and carrying nets. The dried stalks were used in fire-by-friction sets and the feathery seed tails for tinder.

Clematis columbiana (rock clematis) is also essentially a toxic plant. But formulations have been prepared for the treatment of migraine headaches (Tilford 1997). Additionally, liniments made from this plant were once used by physicians for treating nervous disorders and skin eruptions.

Caution The consumption of *C. columbiana* may cause internal bleeding. The entire genus contains strong chemical constituents that can irritate skin and mucous membranes.

Idaho Goldthread (*Coptis occidentalis*)

General Description This is the only species in Idaho. It is a low perennial, evergreen herb with simple, leafless flowering stems. The leaves are divided into three shiny leaflets and are oval in outline, 3-lobed and sharply toothed. The 1-3 whitish flowers have 5-8 sepals and 5-7 narrower petals. The species in common in moist forests at the middle elevations, usually associated with western red cedar, grand fir, and yew. It blooms in early spring, often between patches of melting snow.

Field Notes/Uses The roots of *C. occidentalis* have been used medicinally and contains berberine and coptine. Berberine is a bitter alkaloid that is effective as a digestive stimulant and has strong anti-microbial qualities (see *Berberis* too). Coptine is used as a cholagogue, laxative, astringent, and anti-inflammatory (Tilford 1993). In recent studies of Chinese medicines, *C. sinensis* was shown to be active against HIV, hepatitis B, and several form of influenza, at least in the laboratory setting (Tilford 1997).

Larkspur (*Delphinium*)

General Description Nine species of larkspur occur in Idaho. They are all perennial herbs with tuberous or fibrous roots and erect stems. The leaves are roundish in outline and deeply lobed or divided. The flowers are showy, blue to partly white, containing 5 petal-like sepals with the uppermost prolonged into a spur. There are also 4 petals, 2 partly enclosed by the upper sepals, the lower 2 often hairy and lobed at the tip. They can be found in various habitats, including meadows, thickets, and open woods from the lower to high elevations.

Field Notes/Uses Cattle and horses can contract the usually fatal disease of delphinosis from eating delphiniums (Muenscher 1939). Plants should therefore be regarded as poisonous. Strike (1994) indicates that some *Delphinium* roots were dried, pulverized, mixed with water, and used by the Kawaiisu Indians in California as a salve on swollen limbs.

Mousetail (*Myosurus*)

General Description Any of about 15 species of small, annual, herbaceous (nonwoody) plants constituting the genus *Myosurus*. They occur in the temperate zones of both the Northern and Southern Hemispheres. Mousetails are so named for a long, slender column covered with pistils (female seed-bearing organs) that arises from the center of the flower.

Field Notes/Uses Although tiny flies have been observed visiting mousetails, insects apparently are not necessary to transfer pollen. Stone (1959) noted that tiny mousetails are predominantly self- pollinating. Pollen is shed before the flower opens, when the pistils and stamens are covered by the sepals. Fertilization does not take place until 3 to 10 days later, which ensures that pollen will reach all the pistils that have developed. After the pollen is shed, the flower opens.

Buttercup (*Ranunculus*)

General Description The many species in Idaho are either perennial or occasionally annual herbs with simple to compound leaves. The flowers are solitary or borne in a small inflorescence. The 5 petals are normally yellow or white and have a nectar gland at the base. They can be found in many different habitats from the lower elevations to the alpine zone. The genus name is from the Latin rana for frog and refers to the wet habitat of some species.

Field Notes/Uses All species are more or less poisonous when raw. The leaves and stems should be boiled in several changes of water to remove the poisonous compounds (Kingsbury 1964). The volatile toxin is also rendered harmless by drying (Craighead et al. 1963). The seeds can be parched and ground into meal for bread or pinole. The roots can also be boiled and eaten and were an important part of some Native American diets. A yellow dye can be obtained by crushing and washing the flowers.

Meadow-rue (*Thalictrum*)

General Description The six species in this genus in Idaho are rhizomatous herbs with erect stems. The leaves are 2-4 times branched into ultimate leaflets that are shallowly lobed or toothed, and closely resemble the leaves of columbine (*Aquillegia*). There are no petals, and the 4-5 sepals fall soon after opening. They are found in moist areas at various elevations.

Field Notes/Uses The young leaves of *T. occidentale* (western meadowrue) are said to be edible (Willard 1992). A tea was made from the roots as a cure for colds and venereal disease. The roots dried and powdered can be used as a shampoo. The dried plant of *T. fendleri* (Fendler's meadowrue) was rolled into a cigarette and smoked, or sprinkled on a fire, to treat headaches (Pojar and MacKinnon 1994). Additionally, thalicarpine, a substance used in cancer treatment, has been isolated from *T. pubescens*, *T. revolutum*, and *T. dasycarpum*, species found in eastern North America (Mitchell and Dean 1982).

Carolina Bugbane (*Trautvetteria caroliniensis*)

General Description: This erect, nearly glabrous plant is much branched above and has lower palmately-lobed leaves that are long-petioled. The flowers have no petals. It is found in swamps and along streams.

Field Notes/Uses This species was used by the Bella Coola as a dermatological aid, and as a poultice of the roots was applied to boils.

American Globeflower (*Trollius laxus*)

General Description This plant has showy white flowers with golden centers. The leaves are deeply 5- to 7-parted and occur at the base of the stem. Globeflower is found in wet alpine meadows and along marshy borders of higher elevation streams.

Field Notes/Uses This is a very attractive large-flowered plant that is often associated with *Caltha leptosepala* in cold wet sites at higher elevations. In reality, there are no petals, but rather 5 to 10 petal-like sepals, bright yellow stamens, and green pistils. The undersides of the sepals have a rose-green tinge, which is most easily seen when the flowers are partly closed. At this stage, they appear like small globes, as suggested by the common name.

Buckthorn Family - RHAMNACEAE

Of the approximately 60 genera and 900 species found worldwide, ten genera are native to the United States. Economically they are of little importance, but several species have edible fruits and others are used as ornamentals.

Snowbush, Buckbrush (*Ceanothus*)

General Description The three species found in Idaho are shrubs with leaves that are more or less leathery. One important distinguishing feature is that there are 3 prominent veins originating from near the base of the egg-shaped leaves. The flowers are small, blue or white in color. Look for them on open and dry montane slopes.

Field Notes/Uses The genus has been long recognized as a substitute for commercial black tea and the leaves and flowers could be used to make tea. The seeds can also be used as food. An infusion of the bark may be used as a tonic (Moore 1979). Many species contain saponin which gives the flowers and fruits their soap-like qualities. The flowers when crushed and rubbed in water, will produce a light lather for purposes of washing oneself. Leaves can also be used as a tobacco substitute (Sweet 1976). The long, flexible shoots were used in basketry. The red roots yield a red dye.

Buckthorn (*Frangula/Rhamnus*)

General Description These are shrubs and small trees with flowers that are greenish-yellow and 4 or 5 parted. They are found in wet or moist soils at low elevations.

Field Notes/Uses The genus *Frangula* was previously considered to be only a subgenus of *Rhamnus*, but recent molecular data support its current recognition at the genus level. The two genera can be readily distinguished because *Frangula* has flowers that are usually 5-merous (versus 4 for *Rhamnus*) with petals present (absent in *Rhamnus*), and the style is included in the flower (versus exserted in *Rhamnus*).

Frangula purshiana is used in commercial medicinal preparations (Moore 1979). The bark provides one of the most gentle and best laxatives known. It is usually collected in the fall or spring and dried for a year or more. This plant was also used as a medicine for washing sores and swellings, and treating heart strain, intestinal strains, and biliousness (Pojar and MacKinnon 1994).

A decoction of the leaves of *Rhamnus alnifolia* was used by some Native Americans to soothe rashes caused by poison ivy and oak. The inner bark provided a purgative and a laxative.

Rose Family - ROSACEAE

The Rose family consists of approximately 100 genera and 3,000 species worldwide, with the family being particularly common in Europe, Asia, and North America. About 50 genera occur in the United States. The family is of considerable economic importance because of the edible fruits (e.g., apples, pears, cherries, plums, peaches, apricots, blackberries, raspberries, and strawberries among the important fruits) and many ornamentals.

Serviceberry (*Amelanchier*)

General Description These are shrubs or small trees with simple leaves that are serrate on the terminal half. The white flowers have 5 petals and 5 reflexed sepals, and many stamens. The ovary is inferior and the fruit a pome. The genus name is the French Savoy word for the Medlar (Mespilus germanica), a species that has similar fruits.

Field Notes/Uses All species within the genus Amelanchier produce edible pomes that ripen in late spring and the summer. They were a considered to be a major food for many Native peoples. In fact, some Native Americans intentionally moved their camps to locations where they could be more easily harvested. The pomes may be eaten raw, cooked, or dried. After drying, the pomes can be

Pemmican

Pemmican is a concentrated food used by many early peoples. It is extremely nourishing and does not spoil. To make pemmican, cut meat into strips and dry it until it completely dries and crumbles. It is then ground as fine as possible, usually by pounding. Melted suet (fat) is then poured over the meat, and salt is added for taste as are fresh berries such as currants and serviceberries. The mixture is then kneaded into a paste and packed into containers. In the "old days" the containers were intestines, but you can use plastic-ware. The finished product can be eaten raw, boiled in stews and soups, or fried like sausages.

pounded into loaves or cakes. These in turn may be eaten after softening a piece in water or placing them in soups or stews. Prepared this way, the pomes could be kept for several years. Additionally, the dried pomes could be incorporated into pemmican.

The boiled inner bark was an Native American remedy for treating snowblindness. One drop of strained fluid would be placed in an afflicted eye three times daily (Schofield 1989). It was also used for eardrops and to stop vaginal bleeding. These applications are probably due to the astringency of the plant's tannic acid content.

The wood can be used for arrows, digging sticks, and other useful items. The berry juice makes a purple dye.

Curlleaf Mountain-mahogany (*Cercocarpus ledifolius*)

General Description This is a densely branched evergreen shrub or shrubby tree with gray, furrowed bark. The leaves are leathery, narrowly lance-shaped, and the margins are in-rolled. The flowers are inconspicuous and borne at the tips of branches. The species is a dominant shrub in some areas at low elevations.

Field Notes/Uses The common name, mountain-mahogany, applied to this genus is somewhat misleading. These shrubby trees are not related to true mahogany (*Swietenia*), a valuable cabinetwood of tropical America. The dark reddish brown, mahogany colored hardwood of *Cercocarpus* may have led to this name. Native Americans used the wood for spears, arrow shafts, and digging sticks. The inner brown bark produced a red-purple dye, as did the roots.

A tea to treat colds was prepared by peeling the bark, scraping out its inner layer, and drying and boiling it (Sweet 1976). The dried sap was pulverized and applied to the ears to treat an earache. A decoction of the bark and leaves was used for women's gynecological problems (Strike 1994).

Caution The leaves and seeds of curlleaf mountain-mahogany contain cyanogenic glycosides and should be considered toxic.

Hawthorne (*Crataegus*)

General Description Four species of *Crataegus* occur in Idaho. They are large deciduous shrubs or small trees with thorns. The leaves are toothed or lobed and the white flowers are borne in an open inflorescence. The fruits are small pomes, borne in tremendous quantity, and remaining on the tree all winter. *Crataegus* is a large and varied genus containing many species that readily hybridize.

Field Notes/Uses All species produce edible, albeit mealy fruits which may be eaten raw or cooked in small amounts, or dried and mixed into pemmican. A diet high in hawthorne pomes or drinking hawthorne tea is said to reduce weight (Willard 1992, Moore 1979). The pomes contain a non-toxic heart stimulant and should not be eaten in large amounts or without admixture. Herbal folk medicine considers *Crataegus* a heart tonic (Foster and Duke 1990). Centuries of empirical validation and many scientific studies have shown that the plant is useful in the treatment of hypertension, angina pectoris, and other heart disorders. The pomes also contain Vitamin C. The thorns have many practical uses such as prongs or rakes, lances for blisters, piercing ears, and as fish hooks.

Strawberry (*Fragaria*)

General Description Two species, F. vesca (woodland strawberry) and *F. virginiana* (Virginia strawberry), occur in Idaho. These white-flowered perennial herbs are produced from rootstocks and long runners that root at the nodes. The leaves are clustered at the base of the stem and are divided into three egg-shaped, coarsely toothed leaflets. They can be found in moist, humus-rich, well-drained soils of open forest and forest margins up the subalpine zone. The genus name comes from the Latin *fraga*, the classical name used for the strawberry fruit and referring to its fragrance. The common name, strawberry, comes from the Anglo-Saxon streawberige and refers to the berries "strewing" their runners out over the ground.

Field Notes/Uses Strawberries do not keep well and should be dried for future use if not eaten soon after being picked. Tea made from the green or dried leaves is said to tone up one's appetite (Willard 1992). It may also be a nerve tonic, and was used for bladder and kidney ailments, jaundice, scurvy, diarrhea, and stomach-aches (Foster and Duke 1990). Externally, the leaf tea can also be used as an antiseptic wash for eczema and wounds and as a gargle

for sore throat and mouth ulcers. The plants do contain substantial amounts of Vitamins A and C and sulphur, calcium, potassium, and iron. To remove tartar, rub the berries on your teeth and let the juice sit for a few minutes. Afterwards, brush your teeth thoroughly with baking soda and water.

Avens (*Geum*)

General Description All four species in this genus found in Idaho are perennial herbs that are found in wet, open areas from low elevations to above timberline. Most of the leaves are basal and pinnately divided, whereas the stem leaves are small and less divided. The bell- or saucer-shaped flowers are solitary or borne in an open, few-flowered inflorescence.

Field Notes/Uses A tea from the entire plant of *G. triflorum* (Prairiesmoke) is a general tonic and for severe cough. In addition, its roots can be boiled to produce a weak tea-like liquid to be used as a mouthwash for sore throats and to apply to wounds (Dawson 1985). Moore (1979) suggests that *Geum* spp. is one of the better dysentery remedies - prepare by boiling a tablespoon of chopped root in water and then sip for several hours (2-3 cups daily). The tea is also used for inflammation and irritability of the stomach lining. Scully (1970) indicates that the roots were cooked and eaten by Native Americans.

Oceanspray (*Holodiscus discolor*)

General Description Oceanspray is a deciduous shrub up to 10 feet tall with grayish-red bark. The egg-shaped leaves are shallowly lobed with toothed margins (see photo). The creamy-white colored flowers are small and borne in a diffusely branched inflorescence. Oceanspray is common on rocky slopes and in open forests in the middle elevations.

Field Notes/Uses The small, dry, one-seeded fruits of oceanspray can be eaten raw or cooked. This plant is also used as an astringent, diuretic, tonic and emetic. The stem bark may be decocted and drunk to treat upset stomach, diarrhea, colds, and influenza. As a tonic, it is said to give athletes endurance (Moore 1979). Pojar and MacKinnon (1994) also indicate that some aboriginal peoples steeped the brownish fruiting clusters in boiling water to make an infusion that was drunk for diarrhea. The hardwood can be used for digging sticks.

Ninebark (*Physocarpus*)

General Description Three species are found in Idaho in canyons and slopes in rich soils. They include *P. alternans* (dwarf ninebark), *P. capitatus*

(Pacific ninebark), and *P. malvaceus* (mallow ninebark). In general, they are bushy shrubs from 2-6 feet tall, with grayish or reddish brown, shreddy bark. The leaves are simple and palmately lobed and showy flowers are white and fragrant.

Field Notes/Uses While most Native Americans considered Pacific ninebark as highly poisonous, a tea was made from a stick with the outer bark peeled off. It was used as an emetic and purgative (Pojar and MacKinnon 1994).

Cinquefoil (*Potentilla*)

General Description The many species of cinquefoil in Idaho include perennial, biennial, or annual herbs, and one shrub. The flowers are yellow, white, or, in one case, purple. They can be encountered in various habitat types at all elevations.

Field Notes/Uses The large fleshy, older roots of *P. anserina* (=*Argentina anserina*) (silverweed) can be boiled or roasted and added to soups and stews. Prepared this way they are quite tasty and have a nutty or a parsnip-like texture, but more woody (Peterson 1977). They were a staple among many Native Americans. Today they are seldom harvested, but greatly enjoyed by those who still use them. Silverweed is high in tannins and can be used to tan leather. Other cinquefoils are considered astringent as well (Moore 1979). A tea can be made from the leaves of *P. fruticosa* (=*Pentaphylloides floribunda*) (shrubby cinquefoil) (Weedon 1996). The whole plant or root of *P. arguta* (tall cinquefoil) in a tea or poultice, stops bleeding and has been used on cuts, wounds, and for diarrhea and dysentery (Foster and Duke 1990). A strong tea may still be useful although no longer used as a mouthwash and gargle for sore throats or tonsil inflammations and helps reduce gum inflammation (Moore 1979).

Caution In ancient times these plants were grown for food and medicine. Although there are no reports of toxic reactions from use of this genus, moderation is still advised.

Plum (*Prunus*)

General Description Two species of *Prunus* occur in Idaho. *Prunus emarginata* (bittercherry) is a shrub or small tree with oblong to obovate leaves, which are sometimes hairy. The flowers are white and the fruit bright red. It is found growing on dry or moist sites at the low to middle elevations. *Prunus virginiana* (chokecherry) is a large shrub or tree with elliptic leaves that are finely or sharply serrate. The leaves are also slightly hairy, especially along the midrib. *P. virginiana* has white flowers and dark red to black fruits. It is a common plant at the lower and middle elevations and along streams.

Field Notes/Uses In general, the fruits of both species are sour or bitter when raw, but after cooking the sourness disappears. Native Americans dried the berries whole or in cakes for use in winter. When needed, the dried fruits were soaked in water and then eaten. Lewis and Clark's Expedition members ate chokecherry when other foods were scarce. It seems that after drying, the fruits lose some of their bitterness, resulting in an almost sweet taste.

To make cakes, the ripe fruits are usually ground up, pits and all, and dried in the sun. When needed, the cakes, or portions thereof, can be soaked in water, mixed with flour and sugar and made into a sauce or gravy. This sauce was eagerly traded among some Native Americans such as the Navajo, Shoshone, Arapahoe, and Ute's. The only difficulty we've found in preparing cakes in this manner is that the pits do not grind down nicely into a fine material, leaving larger chunks that could have resulted in broken teeth.

Other uses of the berries included their incorporation into pemmican. They can also be used in making jelly, but because chokecherries are low in natural pectin, it is advisable to add pectin.

The leaves of both species contain toxic amounts of cyanide as do the seeds (pits). Cyanide is highly volatile and the pits can be rendered safe by long-term drying, by boiling in several changes of water, or by dry roasting. Do not eat them in significant amounts even then unless you mix them with larger quantities of other foods. *Prunus* shoots, peeled and split, were used in basketry. The wood was used for various implements, such as digging sticks, arrows, and arrow foreshafts.

Warning Leaves, bark, and seeds of all *Prunus* contain cyanide-producing glycosides. Therefore, eating large quantities of ripe berries with their pits could cause nausea and vomiting. In some instances it could be fatal. Cooking and drying the seeds appears to dispel most of the glycosides and then, the seeds in dried, mashed choke cherries are not as significant a problem. To be safe, it is best to discard the seeds before eating the fruits.

Antelope Bitterbrush (*Purshia tridentata*)

General Description This is a shrub growing up to nine feet tall with alternate, three-lobed leaves and yellow flowers. Considered an excellent browse species for native ungulates, it can be found growing in dry soils at low elevations.

Field Notes/Uses The ripe seeds of Purshia tridentata were boiled and the decoction was drunk as a purgative (Strike 1994). The ripe seed coat produces a violet dye. Old Purshia stumps produced shredding bark that women would peel off, work with their hands to soften, and use as baby diapers. These were sometimes combined with juniper (*Juniperus*) bark (Strike 1994). A boiled leaf decoction was also an important remedy among many Native Americans as a cure for venereal disease.

Rose (*Rosa*)

General Description Six species of *Rosa* occur in Idaho. They are shrubs with prickles with leaves that are pinnately divided into 3-11 leaflets. The large, red to pink flowers are borne singly or a few together. The fruits, called hips, are orange, red, or purplish and urn-shaped.

Field Notes/Uses The hips are edible raw, stewed, candied, or made into preserves. They are high in Vitamin C and also contain Vitamins E, B, and K, beta-carotene, calcium, iron, and phosphorus. There are many other edible parts, besides the fruit. Young rose shoots in spring make an excellent potherb, and the roots and stems can be used to make a tea. The petals may be used in salads. The peeled spring shoots can also be nibbled upon. Almost all parts of the plant have been made into a wash or dressing for cuts or sores to coagulate blood (Willard 1992, Foster and Duke 1990). One of the more common methods is to sprinkle fine shavings of de-barked stems into a washed wound. The petals can be used as a dressing. A poultice of leaves can be used to relieve insect stings. In addition, the young leaves can be washed, cut into small pieces, and dried for a hot tea (Elias and Dykeman 1982).

Raspberry, Blackberry, Thimbleberry (*Rubus*)

General Description Many species occur in Idaho. Our species are deciduous shrubs with arching or trailing stems covered with bristles and prickles. The flowers have white petals and the fruit is a coherent cluster of small, 1-seeded drupes (raspberries, blackberries, dewberries, cloudberries, marionberries). The genus name is derived from the Latin *ruber*, meaning red, in reference to the color of the fruit. This is a large and complicated group taxonomically.

Field Notes/Uses All species produce edible berries. Fossil evidence shows that Rubus species have formed part of the human diet from very early times.

Flowers of all species can be added to salads and can be nibbled upon when hiking. The

Dewberry Juice

"... cover the raw berries with water, crush them, and strain through cheesecloth. Add sweetening to taste. Uncooked juice should be refrigerated and consumed within a day or two."

fresh or dried leaves can be steeped for a tea, alone or in herbal blends. Do not use the wilted or molded foliage, as it may be toxic. The young shoots cut just above ground can be peeled and eaten raw or cooked. A tea from the roots was used to dry runny noses and a tea from the bark was used to stop dysentery (Moore 1979). The plants can also provide a uterine astringent, diuretic, laxative, and mild sedative (Tilford 1993).

Burnet (*Sanguisorba*)

General Description The three species in Idaho. They include *S. canadensis* (Canadian burnet), *S. minor* (small burnet), *S. occidentalis* (western burnet). They are annual, biennial or perennial herbs with pinnately divided leaves. The flowers are clustered in heads and have no petals. The species are found in waste places or moist soils at the lower elevations. The generic name comes from the Latin *sanguis*, meaning blood, and *sorbeo*, meaning to staunch, referring to the herb's ability to stop bleeding.

Field Notes/Uses The young leaves make a good salad plant, tasting somewhat like cucumbers. The leaves can be chopped and blended or mixed with other herbs as a seasoning. The dried flowers and leaves can be prepared as a tea. The roots are very astringent and a decoction was used in the treatment of internal and external bleeding and dysentery. The brew can also be used as a mouthwash for gum problems.

Mountain Ash (*Sorbus*)

General Description Three species of mountain ash are found in Idaho. They are trees or shrubs with deciduous, pinnately compound leaves. The flowers are white to cream-colored and borne in densely branched, flat-topped inflorescences. The fruits are small and berry-like ranging in color from orange to red. They can be found in moist meadows and forest openings at the middle elevations.

Field Notes/Uses The fruits may be eaten raw, cooked, or dried. They are high in Vitamin A and C, and carbohydrates. Unripe berries are very bitter and somewhat unpalatable. The fruits, which are pomes, are commonly processed into jams and jellies. They have high pectin content and jell readily. As a coffee substitute, grind the dried, roasted seeds. The berry juice can be used as a gargle for sore throat and as an antiseptic wash for cuts. Sorbitol, the sugar in the fruit of *Sorbus* is being used commercially for sweetening candies, toothpaste, and other products.

Meadowsweet (*Spiraea*)

General Description Four species of meadowsweet can be found in Idaho. They are small shrubs with deciduous leaves with white to pink flowers that are densely clustered in showy, flat-topped to spike-like inflorescences. The various species can be found in brushy, open slopes and to moist habitats up to timberline.

Field Notes/Uses *Spiraea* is a source of methyl salicylate, similar to the active ingredient in aspirin. Native Americans brewed a tea from the stem, leaves, and flowers of some species to use as a pain reliever (Craighead et al. 1963). The plants are astringent and a poultice made from the leaves and bark was used to treat ulcers, burns, and tumors. The roots were also peeled and boiled until soft, mashed and used as a poultice for burns (Willard 1992). The wiry, branching twigs can be used to make broom-like implements for collecting tubers.

Madder Family - RUBIACEAE

The Bedstraw family consists of approximately 500 genera and 6,000-7,000 species distributed worldwide. About 20 of genera are native to the United States. The family is of economic importance because of coffee, quinine, and many ornamentals.

Bedstraw (*Galium*)

General Description Despite their small flowers, the 10 species of Galium in Idaho are unmistakable. They are annual or perennial herbs with 4-angled stems and whorled leaves. The small, 4-parted flowers are white or greenish and the fruits are smooth or bristly hairy. They can be found in various habitats from the low to higher elevations. *Galium* is from the Greek *gala*, meaning milk, referring to the herbs traditional use as a milk coagulant for making cheese. Rennet (a substance that curdles milk in making cheese and junket) was obtained by blending the herb with an equal amount of salt, covering it with water, and then simmering away half of the fluid.

Field Notes/Uses None of the species of Galium are known to be poisonous. Although *G. aparine* (stickywilly) is the most commonly used species, Tilford (1993) believes, as we do, that all other species can be used similarly. The very young leaves and stems can be used as a potherb. The small hairs on the stems make the plant difficult to swallow raw, boiling or steaming, however, does soften them up. If the stems are too fibrous, use only the leaves. Slow roasted until dark brown and ground, the ripe fruit can be used as a coffee substitute (Peterson 1977).

Medicinally, the plants were used to increase urine flow, stimulate appetite, reduce fevers, and remedy Vitamin C deficiencies. It has diuretic, anti-inflammatory, and astringent qualities, and has been used as a lymphatic tonic (Tilford 1993). A wash made from the plant is said to remove freckles, whereas a cool tea is reported to cool sunburns. Many species of *Galium* contain asperuloside, which produces coumarin, giving it the sweet smell of new-mown hay as the foliage dries. Asperuloside can be converted to prostaglandins (hormonelike compounds that stimulate the uterus and affect the blood vessels), making *Galium* species of great interest to the pharmaceutical industry.

Dried, the foliage of bedstraw has been used as a stuffing for mattresses or as a tinder for starting fires. The roots may yield a red dye, but because the roots are thread-like and produce little dye, collecting enough for a strong dyebath would be fairly laborious.

Willow Family - SALICACEAE

This family has 2-3 genera and over 500 species distributed worldwide. Salix and Populus are native to the United States. The family is of little economic importance, except as a source of ornamentals.

Cottonwoods and Quaking Aspen (*Populus*)

General Description At least three native species of *Populus* occur in Idaho. They include *P. angustifolia* (narrowleaf cottonwood), *P. balsamifera* (balsam poplar), and *P. tremuloides* (quaking aspen). They are trees with sticky, resinous leaf buds, and deciduous leaves. Older trees of some species have gray, rough bark; young bark is smooth and whitish. The flowers are borne in catkins that appear before the leaves. They prefer moist soils and, besides P. tremuloides are usually along streams. They grow rapidly and are planted for

Fire-by-friction

What computers are to the contemporary humans, the ability to start a fire by friction was as important to ancient humans. A quick look at human history reveals fire as the greatest invention of all time.

There are several methods of starting a fire by friction. The drill and hearth method was used by a great many Native Americans. The hearth was a flat piece of wood, usually a softer wood than the drill. It had a small hole reamed in it and a notch leading to the edge of the hearth. The drill was a piece of wood about 2 feet long and a half-inch or less in diameter. Holding the drill vertically and rapidly rotating it between the palms of the hands, embers were formed in the hole. Embers spilled through the notch onto the tinder placed beside the hearth. By cradling the tinder in the hands and gently blowing on the embers, a flame was produced. Depending on craftsmanship, humidity and other factors, a fire could be created in just a few seconds, or never. To help increase the friction, sand grains were sometimes placed in the hole. Also, "points" of ember producing wood were inserted into straight cane-like sticks when appropriate wood could not be found in long straight pieces. The wood these points were made of more efficient in starting a glowing ember. This method of making fire could be accomplished by a single person, or two people taking turns rotating the drill. Another variation used a "bow" to rotate the drill.

Some woods are better than others in starting a fire. A few choice woods in Idaho include sagebrush (Artemisia tridentata), certain species of willow (Salix), juniper (Juniperus), aspen (Populus tremuloides), and cottonwoods (Populus).

quick shade or wind protection. The soft wood of some species is used for veneers, boxes, matches, excelsior, and paper.

Field Notes/Uses The catkins may be eaten raw or boiled in stews and are a source of Vitamin C. The inner bark can also be eaten as a spring tonic, or dried and ground into a flour substitute or extender. The fresh or dried plant can be used in poultices for muscle aches, sprains, or swollen joints (Moore 1979). The primary action of Populus is that of an analgesic, used topically and internally. It contains varying amounts of populin and salicin, compounds related to early forms of aspirin (Tilford 1993). The leaves and bark are most effective parts for tea and aid in diarrhea problems (Moore 1979). The wood makes for an excellent bow and drill fire set. Cottonwoods are considered to be botanical indicators of water and trappers often used aspen as bait in beaver sets.

Willow (*Salix*)

General Description At least 34 species willow are found in Idaho. They are mostly shrubs with numerous stems. Flowers are in catkins that appear before, with, or after the leaves. Willows generally grow along streams or other moist habitats. Willows root easily and occasionally form dense thickets. They are often planted to reduce stream bank erosion.

Field Notes/Uses The young shoots and leaves can be eaten raw. The bitter inner bark can also be eaten raw, although it is better dried and ground into flour substitute or extender. The plant contains salicin which is similar to aspirin and useful as a substitute. Any part of the willow can be used to produce a tea for use as an aspirin replacement for headache and body pain. The highest concentrations of salicin, however, are found in the inner bark. Because it is not nearly as strong as aspirin, you may have to drink quite a bit of it. The leaves have astringent properties that are effective when placed on wounds and cuts (Willard 1992). Bark was chewed as a toothache remedy. Bark, leaves, twigs, and roots produced medicinal teas, powders, washes, and poultices to relieve pain, swelling, infection, bleeding, and many other ailments (Strike 1994). Willows, like the cottonwoods, are botanical indicators of water. The branches of many willow species are very flexible and make them very useful for traps, arrow shafts, and other needs, such as basketry. The bark can also be used as crude cordage.

Willows are an important basketry plant. They were often used as a foundation material and twinning material for twined baskets. Other uses of the wood include framework for dwellings, fish dams and weirs, racks for drying and cooking food, and light hunting bows. Fiber from bark was used for cordage, nets, and clothing. Also, willows root easily due to the large amounts of indole acetic acid (IAA), a plant hormone in their stems. IAA can be extracted in cold water from one inch sections of stem and used to induce rooting of other species for transplanting.

Sandalwood Family - SANTALACEAE

The Sandalwood family is comprised of partially parasitic herbs. There are about 30 genera and 400 species distributed in the tropics and temperate parts of the world. The most common representative in North America is the genus *Comandra*.

Bastard Toadflax (*Comandra umbellata*)

General Description This is a partially parasitic perennial herb with a waxy surface and a rather woody base. The leaves are linear and the flowers are bell-shaped. The fruit is a 1-seeded, berry-like drupe. Bastard toadflax is common and widespread in shrublands up to the subalpine zone. *Comandra* comes from the Greek *kome*, meaning "tufts of hairs," and *aner*, which means "man," in reference to the stamens. The roots are blue when cut.

Field Notes/Uses The mature, brown, urn-shaped fruit of bastard toadflax may be eaten raw, and is best when slightly green. They were popular with Native Americans because of their sweet taste. The berries, however, are rarely found in sufficient quantities for more than a pleasant tidbit. Consuming too many berries may cause nausea. Strike (1994) indicates that a root preparation was used to soothe sore, inflamed eyes.

Saxifrage Family - SAXIFRAGACEAE

There are about 30 genera and 580 species in this family, found chiefly in cooler and temperate regions of the northern hemisphere. About 20 genera are native to the United States, with more species occurring in the western part of the country. Several species are cultivated as ornamentals. The family name means "breaker of rocks" - many members of the genus are found growing in rock crevices. All members of the Saxifrage family are at least somewhat edible.

Alumroot (*Heuchera*)

General Description In general, Heuchera species are perennial herbs with basal leaves. Flowers are small, saucer- to bell-shaped, and greenish, white or pinkish in color. Their various habitats include moist soils and rocky areas up to the alpine zone. The genus is named in honor of Professor J.H. Heucher (1677-1747), custodian of The Botanic Garden in Germany. Many species of alumroot readily hybridize, making identification of some plants difficult.

Field Notes/Uses The leaves of all seven species in Idaho are edible, although they are not choice. They have a sour taste because of the high tannin content. Therefore, the leaves should be boiled or steamed. Since they are rather tough, we found them to be more palatable if chopped and added to soups or salads.

Heuchera is said to be one of the strongest astringents due to their high tannin content. Tilford (1997) indicates that the root of these plants may contain as much as 20 percent of its weight in tannins. Tannins tend to shrink swollen, moist tissues. Therefore, they are also gastrointestinal irritants and have been known to cause kidney and liver failure. Ingestion of the plant should be in moderation. Otherwise, the pounded, dried roots of many species have been used as a poultice that stops bleeding and promotes healing when applied

Dyes, Dyeing, and Mordants

Plants have provided humans with color for uncounted ages. In the last 5,000 to 7,500 years, humans have learned to transfer some of Nature's colors to cloth, paper, wood, leather, and wool. In comparison, natural dyes do not last as long as the synthetics of today, but when you're living primitively, natural dyes are readily available.

Most vegetable dyes are prone to fading and need some added treatment to become color-fast. This process is called mordanting, that is, treating the material to be dyed with other substances that serve to fix the color. Basically, what mordanting and dyeing involves soaking (or boiling or simmering) the material in water with the dissolved mordant. This is dependant on dye source and material to be dyed. After a prescribed time, the material is rinsed and allowed to dry. Dye bath is prepared by soaking the chopped or crushed plant material in water overnight and boiling until the appropriate color is extracted. Plant material is strained out and water is added to make 4-5 gallons of lukewarm dye bath, to which 1 pound of fabric is added. After dyeing and stirring as long as necessary for color desired, the dyed material is passed through a series of rinses, each a little cooler than the previous one, until the rinse water remains clear. After drying, the dyed material is ready for use.

to cuts and sores (Pojar and MacKinnon 1994). The raw root, eaten in small amounts, has been used as a cure for diarrhea (Willard 1992, Weedon 1996). A tea from the roots can also be used as a gargle for sore throats (Moore 1979).

The powdered roots have been used as an antiseptic. The roots of *H. cylindrica* (roundleaf alumroot) were ground to make a fine powder that was rubbed on the skin for rheumatism or sore muscles. Because it was gummy, it clung to the skin.

Alumroots are also commonly used as mordants, substances that make natural dyes colorfast. The alumroot of choice, however, occurs in western deserts near sulphur springs, but satisfactory substitutions probably do exist in Idaho.

Woodland Star (*Lithophragma*)

General Description The three species in Idaho are slender herbaceous plants with stems arising from a perennial, tuberous rootstalk. The leaves are palmately lobed and white to pink flowers are clawed and cleft. They are found in moist or shaded places up to the middle elevations.

Field Notes/Uses I found no recorded uses for Idaho species, but the root of *L. affinis* was chewed by Native Americans in California to treat stomach ailments and colds (Strike 1994).

Saxifrage (*Saxifraga*)

General Description Many species of *Saxifraga* occur in Idaho. In general, they are perennial herbs with basal, alternate, or opposite leaves. The flowers are broadly bell-shaped. The many species occur in various habitats from the foothills to alpine. The generic name is from Latin *saxum* meaning rock and *frangere*, meaning to break. It alludes to the species rocky habitat. Herbalists once used some species in a treatment for "stones" in the urinary tract.

Field Notes/Uses The genus as a whole is regarded as a safe group of plants. The leaves can be used fresh or in stews and are high in Vitamins A and C. Richard Scott (Central Wyoming College) shared with us this favorite back country salad of his:

> "... when backpacking, we always carry a little bottle of vinegar, one of oil, and one of dried or powdered salad dressing. In fact, an excellent salad is based on saxifrage, particularly *S. arguta* (brook saxifrage) leaves with a few alpine sorrel leaves (*Oxyria digyna*), a couple of Indian paintbrush inflorescences (*Castilleja miniata* or *C. rhexifolia*), and a couple columbine flowers (*Aquillea caerulea*)."

In China some species were used in the treatment of nausea and ear infections. In our area, there is little documentation regarding medicinal uses.

Threeleaf Foamflower (*Tiarella trifoliata*)

General Description This is the only species of *Tiarella* in Idaho. It is a perennial herb with mainly basal, simple or trifoliate leaves. The bell-shaped flowers are white to pinkish. It is common in moist forest from mid to high elevations.

Field Notes/Uses No information regarding the uses of threeleaf foamflower in Idaho can be found, but a leaf tea was made from a related species, *T. cordifolia*, and it was also used as a mouthwash. It is considered to be a tonic and diuretic. A root tea was used as a diuretic and diarrhea, and used as a poultice on wounds (Foster and Duke 1990).

Snapdragon Family - SCROPHULARIACEAE

The Snapdragon family has about 220 genera and 3,000 species distributed worldwide. About 40 of the genera are native to the United States. The family is of economic importance because of the cardiac glycosides derived from *Digitalis* (foxglove) and the many fine ornamentals.

Paintbrush (*Castilleja*)

General Description This is a large genus found primarily in western North America that contains many species found in Idaho. The genus is easily recognized, but many species are notoriously difficult to identify. They are perennials with deeply lobed to entire leaves. The flowers are subtended by colorful leaf-like bracts. Some paintbrushes are partial root parasites found in various habitats up to the alpine zone. The genus name honors the Spanish botanist Domingo Castillejo.

Field Notes/Uses Many, if not all of the species have flowers and bracts that can be eaten raw. The seeds of some species were gathered, winnowed, dried, and stored for winter use. In winter they were parched, pounded and eaten dry (Strike 1994). The plants, however, absorb selenium from the soil and so should be taken in moderation. Symptoms in humans of selenium poisoning will vary with the amount and form ingested, but may include difficulty in breathing, excessive urine production, loss of appetite, mental depression, a weak and rapid pulse, blurry vision, digestive upset, and eventually coma and death (Kirk 1975).

Bird's-beak (*Cordylanthus*)

General Description The three *Cordylanthus* species in Idaho are found in dry area at low elevations. They are annual herbs and may be partial root parasites. Leaves are narrow or cut into narrow divisions. The flowers are tubular and 2-lipped with the upper lip incurved and enclosing the style and 2 or 4 of the stamens.

Field Notes/Uses Strike (1994) indicates that *Cordylanthus* was used by Native Americans in California as an emetic.

Toadflax (*Linaria*)

General Description Two species of *Linaria* occur in Idaho, *L. vulgaris* (butter and eggs) and *L. dalmatica* (dalmatian toadflax). Both species are rhizomatous perennial herbs. The flowers are funnel-shaped with 2-lips and a long, narrow, tubular projection (spur) at the base. They are introduced species found in disturbed areas at low elevations.

Field Notes/Uses In folk medicine, a tea made from the leaves of *L. vulgaris* was said to be a laxative and strong diuretic. A "tea" made in milk was used as an insecticide (Foster and Duke 1990). The species has a long history of medicinal use and was once highly regarded as a diuretic for edema. It is seldom used now, but undoubtably merits investigation.

Monkeyflower (*Mimulus*)

General Description Many species of monkeyflower can be found in Idaho. They are annual or perennial herbs with opposite leaves. The flowers flare at the mouth to form 5 lobes, 2 which form the upper lip and 3 lobes that form the lower. The perennial species occur in permanently wet soil, while the annual are found in vernally moist habitats.

Field Notes/Uses The young stems and leaves of *M. guttatus* (seep

monkeyflower) have been used as salad greens. Sometimes, leaves were burned and the ash used a salt (Strike 1994). Weedon (1996) indicates that the young herbage of *Mimulus* species may be eaten in salads, and that they grow bitter with age, but remain edible. For example, the leaves of *M. primuloides* (primrose monkeyflower) were eaten by Native Americans in California and the young plants of *M. moschatus* (musk monkeyflower) were boiled and eaten by some California Natives (Strike 1994).

There are also reports of some Native Americans making a tea from monkey flowers stems, leaves, and flowers as a treatment for kidney or urinary problems and to cure diarrhea (Strike 1994). Monkeyflower leaves and stems were used externally to poultice wounds or internally to reduce fevers. *Mimulus* roots were used to treat fevers, dysentery, and diarrhea and to curtail hemorrhages (Strike 1994). The raw leaves and stems can be applied to burns and wounds as a poultice (Sweet 1976)

Lousewort *(Pedicularis)*

General Description The five species of lousewort in Idaho are perennial, partially parasitic herbs with toothed or pinnately divided leaves. The flowers are subtended by leaf-like bracts and occur in a spike-like inflorescence. The tubular flower is strongly 2 lipped at the mouth. They are found in open, dry or moist habitats, including meadows up to the alpine zone. The genus name comes from the Latin *pediculus* (little louse) alluding to the superstition that livestock that ate the plants would suffer an infestation of lice.

Field Notes/Uses Tilford (1997) indicates that as long as lousewort is not attached to an unpalatable host, the fleshy roots can be prepared and eaten in moderation. He does not, however, describe methods of preparation. He also states that the leaves and stems of some species may be steamed or boiled as potherbs, but this is not recommended.

Moore (1979) describes *Pedicularis* as an effective sedative for children and tranquilizer for adults. The whole stalk, when dried and prepared as a tea, acts as a mild relaxant, quieting anxiety and tension. The fresh or dried plant is a vulnerary for minor injuries, with mild astringent and antiseptic properties.

Warning Because of the sedative nature of these plants, the potential toxic alkaloids, and the host species it may be attached to, ingestion of this plant is not recommended.

Penstemon, Beardstongue *(Penstemon)*

General Description Many species of *Penstemon* occur in Idaho. In general, they are perennial herbs with opposite leaves. The flower is strongly to indistinctly 2-lipped at the mouth with a 2 lobed upper lip and a lower lip with three lobes. There are four anther bearing (fertile) stamens and a single sterile stamen that is often hairy at

225

the tip. The fruit is a many seeded capsule. Penstemons occur in dry or moist meadows or forest openings up into the alpine zone. The genus name is from the Greek *pete*, meaning five, and *stemon*, meaning thread, refering to the slender fifth stamen.

Field Notes/Uses As a topical astringent, pureed or juiced, the plants can be used as a general dressing for minor irritations of the skin (e.g., insect bites) (Tilford 1993). Penstemon oil is a good addition to an all purpose salve (Tilford 1993, Moore 1979).

Lanceleaf Figwort (*Scrophularia lanceolata*)

General Description This is a herbaceous perennial with a thickened root. The stem leaves are opposite and have triangular blades with toothed margins. The 2-lipped flower is yellowish-green and maroon in color. There are four functional stamens and the fifth one is reduced to a small knob on the corolla. This is an uncommon species found in moist meadows and forest openings in the low to middle elevations.

Field Notes/Uses A strong tea made from the plant can be applied to fungal infections of the skin (e.g., athletes foot), and can help eczema, rashes, burns, and hemorrhoids (Moore 1979).

Mullein (*Verbascum*)

General Description Two species of mullein can be found in Idaho. Both are introduced, Eurasian biennial herbs that produce rosettes of basal leaves the first year and a single flowering stem in the second or third year. The white or yellow, saucer-shaped flowers occur in a spike-like inflorescence. The species can be found in disturbed places up to the subalpine zone.

Field Notes/Uses The leaves of *Verbascum thapsus* (mullein) are said to be edible when eaten in small quantities and cooked. Because of their woolly texture, however, we have found the plants to be undesirable.

Native Americans smoked the dried leaves of mullein for asthma and sore throat. A tea from leaves was used to treat colds and the flowers contain oil that has been used for earaches (Moore 1979). The first-year leaves make a soothing decoction for coughs and sore throats. Boil leaves in water for ten minutes, and then strain the liquid through cheesecloth to remove the tiny hairs. The leaves can also be used as poultices applied locally to hemorrhoids, sunburn, and inflammations (Tilford 1993, Willard 1992).

The dried stalks are ideal for use as hand-drills to start fires. The flowers and leaves produce yellow dye; as a toilet paper substitute, the large fresh leaves are choice.

Caution Mullein does contain coumarin and rotenone, two substances that may be toxic in large quantities if ingested. Also, the seeds are not recommended for consumption.

Speedwell (*Veronica*)

General Description Many species of speedwell occur in Idaho. They are annual or perennial herbs with opposite, alternate, or, rarely, whorled leaves. The small, blue, pink, lilac, or white flowers are saucer-shaped and 4-lobed, the upper lobe being the largest. The mature fruit is often necessary for identification. Most species are found in wet soils or shallow waters from low to high elevations in Idaho.

The genus may be named after Saint Veronica, who was said to have wiped the face of Christ on route to his crucifixion. The name may also have arisen to its beginning as a medicinal herb, after a shepherd observed and injured deer heal its wounds by rolling in and eating the herb. The shepherd then reported this to his sick king. The king became well after trying the herb and showered the shepherd with riches.

Field Notes/Uses Although there are many species in Idaho, the species most noted for food and medicine are *V. americana* (American speedwell) and *V. anagallis-aquatica* (water speedwell). Nevertheless, the leaves and stems of all species, when collected during the spring and early summer, can be eaten like watercress, added to salads, or prepared as potherbs. The taste of the various species ranges from spicy to bitter to bland depending on personal taste. The plants also contain moderate amounts of Vitamin C and were once used to prevent scurvy. The leaves and stems can also be steeped as a tea. Care should be taken to avoid plants growing in polluted waters. If desired, the addition of halazone tablets or chlorine bleach to the wash water may kill harmful microbes. After flowering, it is better to boil the plants to eliminate the bitterness.

Medicinally, the plants were used mainly as an expectorant for respiratory problems. Infusions were also used in hair conditioning rinses and skin cleaning herbal steams, and as an ingredient in massage oils and ointments that were added to baths for a soothing soak. Pojar and MacKinnon (1994) indicate that *V. americana* has been used for centuries to treat urinary and kidney complaints and as a blood purifier. The leaf juice from *V. serpyllifolia* (thymeleaf speedwell) has been used for earaches, its leaves were poulticed for boils, and a tea was used for chills and coughs (Foster and Duke 1990).

SOLANACEAE (Nightshade Family)

There are about 85 genera and 2,300 species in the Nightshade family. About 13 genera are indigenous to the United States. The family is noted for its edible, poisonous, and medicinal plants. Some of the economically important plants include tomato, potato, chili, tobacco, and eggplant. Medicinally, two alkaloids, atropine (belladonna) and scopolamine which can be deadly in large amounts are obtained from this family. Eye doctors today use atropine in very exact small, doses as an effective dilator when examining eyes.

Datura (*Datura*)

General Description Two species of Datura may be encountered in Idaho, *D. stramonium* (jimsonweed) and *D. wrightii* (sacred thorn-apple). Each has an ill-smelling odor. In all *Datura* species, the funnel-shaped corollas may be tinged to varying degrees of violet. The fruits are a leathery or woody capsule, with prickles. They are found in disturbed areas at the lower elevations and foothills.

Field Notes/Uses Extracts from these plants are narcotic and, when improperly prepared, are very lethal. *Datura* is also known as "hell's bells" among drug users because of the popular misconception that the plants provide worthwhile hallucinogenic experiences. Unfortunately, those who experimented with *Datura* as a drug have died as a result. Those that did survive realized that it was not worth the try. The narcotic properties have been known since before recorded human history and they once figured importantly in the religious ceremonies of many Native Americans.

Although potentially deadly, the plants do have valuable medicinal values. There are several alkaloids including atropine and hyoscyamine that have been used in scientifically refined antispasmodic drugs. Datura also contains scopolamine, an antivertigo compound commonly used to treat motion sickness and other conditions involving disequilibrium (Tilford 1997). The concentrations of these compounds vary widely among plant parts, species, and localities, making Datura highly unreliable for internal use, even in trained hands.

Warning These plants should never be taken internally for any reason. They contain dangerous alkaloid compounds that could easily kill a person.

Black Henbane (*Hyoscyamus niger*)

General Description This plant grows up to 2 feet tall, and has large, pale green, oval leaves with deeply toothed edges. Tiny hairs cover the stem and leaves. The flowers are bell-shaped and have mustard-yellow petals with purplish-brown throats and veins. The seeds are enclosed in $\frac{1}{2}$ inch long capsules.

Field Notes/Uses This is a native plant to western Asia and southern Europe, but is now found across much of western and central Europe, and North and South America. It is cultivated for therapeutic use in parts of Europe, including England, and in North America. The leaves and flowers are picked just after the plant has flowered, in the first year for the annual variety and in the second year for the biennial.

A traditional witches' brew ingredient, henbane has suffered from a deservedly sinister reputation ever since ancient times. The narcotic alkaloids hyoscyamine, scopolamine, and atropine are derived from this ugly, foul-smelling weed. All parts of the plant are poisonous, and if eaten, even small amounts cause anything from dizziness to delirium. Too much brings slow and painful death. In past times, henbane served as a sedative to ease pain and spasms, but the determination of a safe dose has always been a tricky business, and for long periods the drug seems to have been left alone by medical practitioners.

Matrimony Vine (*Lycium barbarum*)

General Description This is an introduced species. In general, *Lycium* are shrubby plants with entire to minutely toothed leaves that sometimes grow in bundles. The purple to greenish-purple flowers are borne solitary, or in small clusters in the leaf axils. The roundish berries are fleshy to dry, depending on the species.

Field Notes/Uses The fruit of matrimony vine (and other *Lycium* species) are edible raw or cooked and has a mild sweet liquorice flavor. Only the fully ripe fruits should be eaten, unripe berries could be poisonous. As a food, dried wolfberries are traditionally cooked before consumption. Kirk (1970) suggests all *Lycium* species berries (commonly known as wolfberries) are edible; some better tasting than others.

Young wolfberry shoots and leaves are also grown commercially as a leaf vegetable. In the West, dried wolfberries are also eaten hand-to-mouth as a snack, in the manner of raisins or other dried fruit. Their taste has an accent of tomato and is similar to that of dates, dried cranberries or raisins; though drier, more tart, and less sweet and with an herbal scent. Dried wolfberries are also used frequently in raw food diets.

Coyote Tobacco (*Nicotiana attenuata*)

General Description This annual plant grows up to 40 inches tall, and has glabrous to glandular pubescent stems. The leaves are large, ovate to ovate lanceolate in shape, and the upper leaves are narrowed. Flowers occur in a terminal raceme, and the calyx is 5-cleft. Corolla is white, tubular or funnel-shaped. The flowers close in the sun. Coyote tobacco grows in disturbed places and flowers from May to October.

Field Notes/Uses All species of *Nicotiana* contain the highly toxic alkaloid nicotine. An effective insecticide against aphids can be prepared by steeping tobacco leaves in water and spraying the solution on affected parts of the plant.

Ground Cherry (*Physalis*)

General Description Three species of *Physalis* can be found in Idaho. They are annual or perennial herbs with alternate, entire to coarsely toothed leaves. The flowers are either solitary in the axils of the leaves or in clusters of 2-5. The calyx is 5-lobed and becomes enlarged and inflated in fruit -- a many seeded berry. The genus name is from the Greek meaning bladder, referring to the inflated calyx. The plants are found in the waste places and fields at the lower elevations.

Field Notes/Uses The berries can be eaten raw or cooked and taste best when fully ripe. There is a tendency for the fruits to fall off the plants before maturation. In that case, simply gather the fruits and allow them to ripen in the husk. They will keep for a couple of weeks in this condition. The fruits have been used in preserves and pies. The unripe fruit of some species are considered to be poisonous when eaten in large amounts. If the fruits are bitter, sour, or otherwise strongly flavored, they are unripe. In any case, let them ripen until soft and sweet.

Nightshade (*Solanum*)

General Description This is a highly diverse genus comprising more than a thousand species worldwide. In Idaho there are about ten species of *Solanum*. In general, nightshades are annual or perennial herbs with flowers that resemble those of tomato or potato plants. The fruit is a many-seeded berry, surrounded in part by the persistent calyx. The various nightshades can be found from moist, open habitats to disturbed, waste areas at the lower elevations. Solanum probably comes from the Latin "*solamen*," which means

quieting, referring to the sedative properties of some species.

Field Notes/Uses The most poisonous part of the plant is the unripe fruit, but the stems, leaves, and roots are also dangerous. In fact, the alkaloid content in the plants decreases in this order: unripe fruit-leaves-stems-ripe fruit (Schofield 1989). Solanine is the predominant glyco-alkaloid, but others may be present. Solanine is highly toxic and can cause death, but the degree of toxicity varies among and within species. Craighead et al. (1963) suggest that cooking destroys the solanine, and making the ripe fruit edible.

The berries of *S. nigrum* (black nightshade) were formerly used as a diuretic (Foster and Duke 1990). Native Americans of the Southwest used the crushed berries to curdle milk for making cheese and they have also been used in various preparations for sore throat and toothaches.

Solanum dulcamara (climbing nightshade), when used correctly and in appropriate dosages, is said to be useful in the treatment of skin disorders, rheumatism, and bronchitis. Recent studies have shown that this plant possesses anti-cancer qualities (Tilford 1997)

Caution There are references listing the berries of some species of *Solanum* as edible. However, it is our recommendation that none of the plants that occur in Idaho be consumed in any manner.

Elm Family - ULMACEAE

Members of this family are trees or shrubs with simple, alternate leaves. The flowers lack petals and the fruit is a samara, nut, or drupe. There are approximately 15 genera and 200 species in this family distributed through the temperate and subtropical regions of the Northern Hemisphere. *Ulmus*, *Planera*, and *Celtis* are native to the United States. The family is of economic importance as a source of wood.

Hackberry (*Celtis laevigata*)

General Description This is a small tree or shrub with leaves that are ovate to lanceolate in shape, with entire to serrate edges. The fruit is a drupe, and the plants can usually be found growing along streams or on dry canyon slopes at the lower elevations.

Field Notes/Uses The small orange, red, or yellow fruits are edible raw and have a sweet taste to them. The entire fruits can also be dried and then ground into a flour.

Nettle Family - URTICACEAE

There are about 45 genera and 550 species found in the Nettle family. Six of the genera are native to the United States and are of little economic importance.

Pennsylvania Pellitory (*Parietaria pensylvanica*)

General Description Pellitory, in places called cucumber weed, is recognized by its long-tapering alternate leaves, the lack of stinging hairs on the stems, and the small axillary clusters of green flowers. There are several flowers crowded into small axillary clusters, some of the flowers perfect, some male only, some female only, all often in the same cluster on the same plant, each flower green, is surpassed by bracts.

Field Notes/Uses This inconspicuous little plant is easy to overlook. Unlike some other members of the Nettle family, pellitory lacks stinging hairs and its foliage is harmless.

This is one of those forgotten edibles that never seem to be included in any edible plant book. Too bad, as the leaves have a cucumber-like taste. The older leaves are a bit fuzzy and would require a little cooking. However, when young, the leaves are most desirable for eating.

Stinging Nettle (*Urtica dioica*)

General Description This is an annual or perennial herb with stinging hairs. The flowers are numerous, small, and clustered on drooping branches at the base of the leaves. Stinging nettles can be found along roadsides, streams, in moist areas and waste places in the low to middle elevations. Stinging nettle is an indicator of good soil conditions. Many people consider stinging nettles, for obvious reasons, as obnoxious weeds. Recent taxonomic revisions have consolidated various *Urtica* species into a single species.

Field Notes/Uses One of the first things a person learns about stinging nettles is their stinging effect. The intense burning and itching or stinging of the skin may persist for a length of time. If you look closely at the hairs, you will see a hypodermic mechanism consisting of a very fine capillary tube with a bladder-like base that is filled with chemical irritant. When brushed against, a minute spherical tip breaks off uncovering a very sharp-pointed tip

that easily penetrates the skin. The chemical is forced into the skin through the tube as the hair bends and constricts the bladder-like base (Krochmal 1973, Muenscher 1940). Therefore, stinging nettles should be collected with gloves.

The young stems and leaves of stinging nettle are edible after boiling, and are very delicious as a spinach substitute. Boiling the leaves destroys the formic acid found in the hairs. The leaves are high in vitamins A, C, and D, the latter of which is rare in plants. The roots are also edible after they have been roasted. A tea made from the leaves is said to have astringent and diuretic qualities, and has been used for internal bleeding and nosebleeds (Moore 1979). Native Americans learned of stinging nettle as a food from early European travelers and settlers, and possibly from Chinese immigrants (Kuhnlein and Turner 1991).

The older stems become fibrous, which reduces their edible qualities, but allows them to be used to produce strong cordage. The older leaves also contain cystoliths that can irritate the kidneys. A yellow dye may be obtained by boiling the roots.

A tea brewed from *Urtica* was said to relieve chest colds and internal pains (Strike 1994). As a poultice, *Urtica* was used for headaches.

Valerian Family - VALERIANACEAE

The Valerian family has about 13 genera and 400 species. Three of the genera are native to the United States. In general, the family is of no economic importance.

Valerian (*Valeriana*)

General Description The six species of valerian in Idaho are perennial herbs with aromatic (actually ill-smelling) roots. The stem leaves are opposite and pinnately compound and the flowers have three stamens. They can be found in open forests and meadows to timberline and above. The genus name comes from the Latin *valere* (to be strong) and refers to medicinal qualities of the plant.

Field Notes/Uses The plants commonly contain alkaloids chatinine and valerine. They are known to act upon the central nervous system as depressants and are prescribed to calm nerves and relieve insomnia. The plants were used by physicians since at least the 9th century. Extracts were used as nerve tonics and may rival the relaxing properties of opium. Valerian was one of 72 ingredients Mithridates, King of Pontus, compounded as an antidote to poison, using poisoned slaves as test subjects.

The roots and leaves of *V. edulis* (edible valerian) can be collected, steamed for a day or two to remove the disagreeable odor, and then used in soups as a potato substitute. However, the taste does get a little getting used to and may remain somewhat unpalatable. We found that the steamed roots are better if dried, ground into flour, and then added to other flours. The other species could probably also be used in an emergency, but they do not have the large taproot of V. edulis.

Valerian, in general, are considered best known for their calming qualities. They have been used for more than a hundred years as a remedy for anxiety, muscle tension, and insomnia. The plants contain valepotriates, which is a known herbal calmative, antispasmodic, and nerve tonic, and is used for hypochondria, nervous headaches, irritability, and insomnia (Tilford 1993). Research has confirmed that teas, tinctures, or extracts of this plant are a central nervous system depressant and a sedative for agitation (Foster and Duke 1990).

Verbena Family - VERBENACEAE

Approximately 75 genera and 3,000 species occur worldwide, of which 14 genera are native to the United States. The family is of economic importance because of the highly prized teak wood (*Tectona grandis*) and a number of other ornamentals.

Verbena (*Verbena*)

General Description Three species can be found in Idaho. Members of this genus are perennials with opposite, toothed leaves. The tubular flowers with flaring lobes each have a subtending bract. The fruit is a cluster of 4 nutlets.

Field Notes/Uses The seeds of *V. hastata* (blue vervain) may be gathered, roasted, and ground into bitter tasting flour. Leaching the flour may remove the bitter taste. A tea from boiled leaves can be used for a stomachache, and a tea from roots was used to clear cloudy urine (Willard 1992, Foster and Duke 1990). Moore (1979) says that the plant is used as a sedative, diaphoretic, bitter tonic, and mild coagulant. It promotes sweating, relaxes and soothes, settles the stomach, and gives an overall feeling of relaxed well-being.

Violet Family - VIOLACEAE

The Violet family has approximately 16 genera and 850 species distributed worldwide. Two genera are native to the United States. The family is of little economic importance other than a source of ornamentals as many species of Viola are cultivated. Viola is the classical name for violets.

Violet (*Viola*)

General Description Nineteen species of violets occur in Idaho. In general, they are low-growing, perennial or annual herbs. The leaves are spade-shaped and basal. The flowers occur singly on the ends of stems and have five petals. There are 2 upper and 2 lateral petals, and 1 lower petal that is prolonged into a nectar holding pouch at the base of the flower. Most species also have small, self-fertilizing flowers that do not open. Violets can be found in meadows and open forests from the foothills to above timberline.

Field Notes/Uses The leaves, buds, and flowers of possibly all species are edible raw or cooked, with some being more palatable than others; the leaves make a good tea. Adding the leaves to soups make them thicker. Violets are high in Vitamin C and beta-carotene. Collect the plants by leaving the roots intact. Since many species reproduce vegetatively, you will probably not inhibit next year's growth significantly. Many naturalists indicate that all violets are safe for consumption, but there are some experts that insist some yellow species may be somewhat purgative. All species do have a tendency to be slightly laxative, so proceed slowly. The flowers have also been candied or made into jellies and jams.

Violet salve can be made by simmering the entire herb in lard. It was a famous remedy for skin inflammations and abrasions. Violets are also emollients (soften the skin) and are an excellent ingredient in lotions such as night creams. Flowers and leaves of some species have been used in various herbal remedies as poultices and laxatives and to relieve cough and lung congestion (Pojar and MacKinnon 1994).

Mistletoe Family - VISCACEAE

These are parasitic plants found on the branches of a variety of trees such as oaks, alders, conifers, and cottonwoods. Leaves are opposite and somewhat leathery. The flowers are small, comprised of 2-4 tepals (petals and sepals not differentiated) inserted on a cup-like receptacle. The fruit is a drupe or berry. There are about 11 genera and 450 species in this family distributed worldwide. In the United States, *Arceuthobium* and *Phoradendron* are native.

Dwarf Mistletoe (*Arceuthobium*)

General Description Five species of dwarf mistletoe occur in Idaho. They grow as parasites on coniferous trees and have opposite leaves. The flowers are found in the axils of the leaves. The pulp of the berry is mucilaginous which glues the seed to whatever it touches (to eventually germinate on the trees). These mistletoes are found on a variety of conifers. The genus name is from the Greek *arkeuthos*, meaning juniper, and *bios*, meaning life, as the plant is often a parasite on junipers.

Field Notes/Uses Strike (1994) indicates that A. campylopodium (pine mistletoe) mixed with the pitch of *Pinus edulis*, was used to cure coughs and colds and to relieve the associated aches. Additionally, the smoke from burning mistletoe relieved colds and coughs and a decoction was used as a contraceptive.

Caltrop Family - ZYGOPHYLLACEAE

The Caltrop family is comprised of herbs and shrubs, and rarely, trees that exhibit a xerophytic or halophytic life history. The leaves are usually opposite and pinnately compound. There are approximately 30 genera and 250 species distributed primarily in tropical and subtropical areas. Six genera are native to the United States. Guaiacum wood and lignum vitae (a valuable heavy hardwood and wood resin used in some chemical tests) are the important economic products in this family.

Goathead, Puncture Vine (*Tribulus terrestris*)

General Description This is an introduced annual plant with trailing, hairy stems, and an extensive root system. The leaves are opposite and pinnate with four to eight pairs of leaflets. The flowers are yellow and are borne singly in the leaf axils. The fruit is hard, consisting of five spiny nutlets or burrs that break apart into five "tack-like" sections upon maturity. These burs may injure livestock and are the bane of bicyclists. The plant is found in disturbed and waste places at the lower elevations.

Field Notes/Uses Puncture vine has a 5,000 year history of medicinal uses, particularly in China and India. It was used for boosting the hormone production in men and women, and for urinary tract problems, itchy skin, and blood purification. The stems of the plant are considered to be astringent and act upon the mucous membrane of the urinary tract. A tea from the aboveground part of the plant is said to be good for arthritis.

Moore (1979) indicates that studies have been conducted on the seeds that show they are useful in the early treatment of elevated blood fats and cholesterol. It apparently helps prevent or lessen the severity of arteriosclerosis and atherosclerosis. Moreover, he says that the plant is useful in treating mild hypertension, contributing to a slower, stronger, more well-defined heart function, with greater relaxation between contractions and a lowering of the diastolic pressure.

Syrian Beancaper (*Zygophyllum fabago*)

General Description This is a large, introduced perennial forb with thick stems that become woody over time. The plant can grow up to 3 feet tall and is bushy. The fleshy leaves are thick, smooth, hairless, and oblong in shape. They grow in pairs on stems that are opposite one another on the main branches. The flower stalks grow from where the leaf stems meet the main branches. Five petals are white to cream colored with a pinkish tinge. Each of the flowers has ten orange stamens extending beyond the petals. Seeds are produced in five chambered pods with one seed in each chamber. The pods are oblong and cylinder shaped with five sides.

Field Notes/Uses Syrian bean-caper is an invasive plant that dominates dry habitats such as rangelands and grasslands. It reduces native biodiversity by forming dense patches that compete with native plants for water and space. Livestock will not eat Syrian bean-caper. Syrian bean-caper is difficult to control with herbicides and is drought tolerant. People plant Syrian bean-caper as a medicinal herb and caper substitute. Syrian bean-caper is usually spread by seed, and can generate new plants from pieces of its roots. It can also be spread by farm machinery.

Section IV - Angiosperms: Monocots

Monocots are distinguished by a number of features. In monocots, the embryos of the seeds have only one cotyledon (seed-leaf). The plant leaves usually have parallel veins. The vascular bundles of the stems are irregularly arranged, and the cambium is lacking. The flower parts are arranged in threes or sixes, never in fives (or fours). Monocots are usually herbs, rarely shrubby.

Arrowhead Family - ALISMATACEAE

Thirteen genera and 90 species in this family are found worldwide, of which five genera are native to the United States. Members of this family are aquatic and marsh perennial herbs. Their long-stalked leaves and even longer flowering stems are well adapted to wetland habitats. The flowers are comprised of 3 green sepals, 3 white petals, numerous stamens, and many pistils. This family is of little economic importance. There are three genera in Idaho.

Water Plantain (*Alisma*)

General Description Two species of water plantain occur in Idaho: *A. triviale* (northern water plantain) and *A. gramineum* (narrowleaf water plantain). They are perennial plants from fleshy, bulb-like stems. The basal leaves are long stalked and egg-shaped, and the flowers are white. Water plantain is usually found in marshes and ponds at lower elevations.

Field Notes/Uses The starchy, bulbous bases of water plantain are edible as a starchy vegetable (potato) after drying. Drying is said to remove the strong flavor. Alisma has a long history of use in Chinese medicine and is

mentioned in texts dating back to about 200 A.D. It was also used by early herbalists as a diuretic and by the Cherokee Indians for application to sores, wounds, and bruises. It is described as a sweet, cooling herb that lowers blood pressure, cholesterol and blood sugar levels. The root was also used as a diuretic in the treatment of dysuria, edema, distention, diarrhea, and other ailments (Foster and Duke 1990). Water plantain also furnishes food for water birds and muskrats.

Arrowhead (*Sagittaria*)

General Description Also known as wapato and duck potato, arrowheads are aquatic plants found in a variety of shallow wetland habitats such as streams, ponds, marshes, and ditches. *Sagitta* is Latin for "arrow" referring to the shape of the leaves. The sap of the plants is milky. Flowers have three showy petals and are attached to the stalk in whorls of three. There are three species of arrowhead in Idaho.

Field Notes/Uses All species produce starchy, white tubers that can be roasted or boiled, and then eaten. An important source of carbohydrates, the tubers contain a milky juice with a bitter flavor that is destroyed by heat. The Lewis and Clark Expedition is said to have utilized these tubers extensively while exploring the Columbia River region. To many Native American tribes, arrowhead was a primary vegetable. The small tubers are located at the ends of the long underwater rhizomes, perhaps a meter or more from the plant. They can be carefully removed without pulling up the whole plant.

To collect the tubers, you can use your hands, a forked stick, or if the water is deep, your feet. Wade into the pond where the plants are growing and feel around for the tubers in the mud with your toes. They should feel like round lumps varying in size from peanut to potato. After dislodging them, the tubers usually float to the surface for easy harvesting. They are best developed in late summer or autumn. Boil or bake them like a potato to remove the poisonous properties, then peel and eat them. They can also be dried for future use. The tubers were used in a tea for indigestion, and in a poultice for wounds and sores (Foster and Duke 1990). The leaves and stem contain alkaloids and may be poisonous.

Arum Family - ARACEAE

These are fleshy perennials with basal leaves. The flowers are borne in dense fleshy spikes (spadix), subtended or enclosed by a foliaceous or colored bract (spathe). There are more than 115 genera and 2,000 species that are found in shady, damp or wet places. The family is well represented in the tropics. Members of this family include Calla Lily, Philodendron, and Dieffenbachia which are grown as ornamentals.

Sweet Flag (*Acorus americanus*)

General Description The leaves of sweet flag are sword shaped, and when bruised, the whole plant is fragrant. This plant is found mainly in the northern part of the State and is widely established in the eastern part of the country.

Field Notes/Uses The drug calamus is obtained from the rootstalk (Davis 1952).

American Skunkcabbage (*Lysichiton americanus*)

General Description American skunkcabbage is found in marshes and wet woods at lower elevations in the northern part of the state. The genus name is from the Greek *lysis*, meaning a loosening, and *chiton*, meaning tunic pertaining to the large spathe.

Field Notes/Uses The plant contains calcium oxalate in all parts. Cooking the plant renders it edible. Native Americans roasted or dried the root and made flour from the starch. The young green leaves can be eaten after

boiling in several changes of water. Drying or heating removes the stinging properties. This plant is related to Taro, a staple food of the Polynesians (Craighead et al. 1963).

Sedge Family - CYPERACEAE

There are 90 genera and 4,000 species worldwide, particularly in the cool temperate and sub-arctic regions. Twenty-four genera are native to the United States. In Idaho, sedges are a large family of grass-like plants often found growing in wet places. Sedges make up a large proportion of the plants found at higher elevations. Most of the species have triangular stalks with the flowers arranged in spikelets. The family is of little economic importance, although several members are edible and medicinal. Sedge family members, however, should not be used medicinally without medical supervision since their properties and effectiveness are variable.

Sedge (*Carex*)

General Description Numerous species within this genus occur in Idaho. They are normally difficult to identify in the field. A hand lens and mature fruit are often necessary for positive identification of most species. In general, they are perennial grasslike herbs with creeping rhizomes, short rootstalks, or fibrous roots, with 3-sided stems, and the leaves are 3-ranked. Sedges occur in a great variety of habitats from the foothills to the alpine zone.

Field Notes/Uses The young shoots and tender leaf bases of almost all species are sweet and furnish a tasty nibble (Craighead et al. 1963). The fruits are also edible. *Carex* is a widespread genus with more than 1,000 species, making it available as an important emergency food in many parts of the country. *Carex* roots were used extensively as basketry material, particularly in coiled baskets. Some Native Americans would spend considerable time untangling roots in order to remove them in one piece. In fact, some areas were cleared of other plants to allow Sedge to grow long and untangled (Spike 1994).

Nut-grass (Cyperus)

General Description Members of this genus are annual or perennial and grasslike. Leaves are basal, three- ranked, and the stems triangular in cross-section. The spikelets are clustered in ball-shaped heads and the inflorescence consists of numerous heads borne on stalks radiating from the top of the stem and subtended by long, leaf-like bracts. The various species occur in wet, open soils along river banks, and the margins of lakes and ponds.

Field Notes/Uses Of the 10 species of nut-grass that occur in Idaho, the rhizome of *C. esculentus* (chufa flatsedge) bears small, nut-like, underground tubers that are edible. These may be eaten raw, boiled, dried and ground into flour, or roasted to a dark brown

Nut-grass Drink

This is a Spanish recipe for a refreshing drink. Soak about ½ pound of tubers in water for 48 hours. Then mash the tubers, add 1 quart of water and 1/3 pound of sugar. Then strain the liquid through a sieve and serve as a drink (Fernald and Kinsey 1958).

and ground into coffee (Peterson 1977). The species was so valued in ancient times that its tubers were placed in Egyptian tombs dating back to more than 2,000 years B.C. Other species may also have tubers or tuber-like structures. Volatile oils and astringent substances are found in a number of species that are used in perfumery and as remedies for digestive problems.

Threeway Sedge (Dulichium arundinaceum)

General Description One species, *D. arundinaceum*, occurs in Idaho. It is a perennial herb found with rounded stems arising from deep-seated rhizomes. It can be found growing in wet meadows and along the margins of ponds and streams at low to middle elevations.

Field Notes/Uses The rhizome is probably edible, but it is usually found among other plants of higher food value (Weedon 1996).

Spikerush (*Eleocharis*)

General Description Spikerushes are annual or perennial plants that have needle-shaped or flattened stems with leaves reduced to sheaths or sheathing scales. Each flower is subtended by a papery scale and consists of an ovary with 3 stamens and the petals and sepals reduced to bristles at the base of the ovary or completely lacking. The flowers are spirally arranged in a single, congested spike at the tip of the stem. The fruit is an achene that is 2- or 3-sided in cross section. The thickened base of the style persists as a small projection (tubercle) on the top of the achene. Most species mature fairly late in the growing season, and mature fruit is necessary for positive identification.

Field Notes/Uses These sedges can be distinguished from other sedges by the presence of a tubercle (a small bump, like a nipple) on the achene (or fruit) and the fact that the stems grow in groups and are usually tufted. *Eleocharis* derives from the Greek words *heleios* (marsh) and *charis* (joy, beauty or grace). Hence the name means "marsh beauty." Common spikerush is an invaluable food source to waterfowl. Small mammals seek shelter in dense stands of this plant but do not utilize it for food.

The author has used the seeds (actually achenes) of *E. palustris* (common spikerush) as flour after grinding. Common spikerush is a variable species and is the most common member of the genus in our area, occurring in many wet, or vernally wet, open habitats in the valley and montane zones. It is a widespread species in temperate and boreal regions of the Northern Hemisphere. Chances are other species may be suitable. The stems of almost any spikerush can be used in weaving.

Cotton Grass (*Eriophorum*)

General Description The four species of cotton grass in Idaho are perennials with solid, triangular or round stems from rhizomes. The lower leaves are grasslike. The flowers are spirally arranged in spikes that resemble tufts of cotton at maturity. *Eriophorum* translates as "wool" (*erion*) and "bearing" (*phoros*). The species can be found in cold swamps and bogs from mid to high elevations.

Field Notes/Uses The pinkish bases of these

plants can be collected in the early summer and eaten raw or added to soups. Dug in the early spring or fall, the corms were prepared by pouring boiling water over them to remove the thick black outer covering removed. The corms were then eaten raw or boiled. A decoction from the rootstock was used in treating colds and coughs, but its use was not widespread (Schofield 1989). The woolly down from the flowers was spun as fiber, although it was inferior to cotton. However, the fiber can be used as candlewicks.

Lipocarpha (*Lipocarpha*)

General Description The genus was formally known as *Hemicarpha*. The two species in Idaho, *L. aristulata* (awned halfchaff sedge) and *L. micrantha* (smallflower hemicarpha) are small annuals with three-angled leafy stems found in damp locations.

Field Notes/Uses The pollen and seeds are edible (Weedon 1996).

Bulrush (*Scirpus*)

General Description The many species of bulrush in Idaho are perennials with grasslike or scale-like leaves. The various species can be found in marshy areas around lakes and ponds, and in other moist or wet areas from low to high elevations.

Field Notes/Uses The edible rhizomes of all species are quite starchy. They may be eaten raw or baked, dried, or ground into flour. As Olsen (1990) describes:

> "...the young shoots just protruding from the mud are a delicacy raw or cooked. Furthermore, one can harvest them by wading into the water and feeling down along the plant until you come across the last shoot in a string of shoots that protrudes above the water. You then push your hands into the mud until the lateral rootstalk is encountered. By feeling along the rootstalk in a direction leading away from the last shoot, one can find a protruding bulb from which the new shoot is starting. This is easily snapped off and is edible on the spot."

The young roots crushed and boiled also yields sweet tasting syrup. The pollen may be gathered and pressed into cakes and baked. The seeds may be used whole, parched, ground, or in mush. The stem bases may be eaten raw and are good for quenching thirst.

Scirpus stems, roots, and leaves can be used as the foundation and twining material in twined baskets. They were used extensively by Native Americans for making cordage, sandals, baskets, and mats. Waterproof "water bottles" can be made from the baskets by coating the inside with asphaltum.

The stems were used for sleeping mats, padding, thatching dwellings, skirts, and sandals. Duck-shaped decoys were also made when hunting.

Leaves of *S. acutus* (also known as *Schoenoplectus acutus*) were used to poultice wounds and burns (Strike 1994).

Iris Family - IRIDACEAE

There are about 70 genera and 1,500 species in this family worldwide, with the chief center of distribution in South America and tropical America. Five genera are native to the United States. In Idaho, members of this family are perennials with narrow, grasslike leaves and thick rhizomes or fibrous roots. The flowers have three petals and three petal-like sepals joined on top of the ovary. The family is of economic importance as a source of ornamentals and saffron dye.

Iris (*Iris*)

General Description Of the three species of *Iris* in Idaho (2 native/1 non-native), a person is likely to encounter *I. missouriensis* (Rocky Mountain iris). The stems of this plant are about two feet tall with several grass-like leaves. The flowers are showy with three drooping blue sepals and three petals that are slightly smaller than the sepals. Rocky Mountain iris can be found in meadows, wet or moist areas at low to mid elevations. Iris is the Greek word for rainbow.

Field Notes/Uses Members of the genus *Iris* contain irisin, an acrid resin concentrated mainly in the rhizomes, and present in the foliage and flowers. People who raise irises sometimes develop a skin rash from handling the rhizomes. Cattle have died as a result of eating relatively large quantities of the plants. The rootstock produces a burning sensation when chewed. If eaten in quantity, irises will cause diarrhea and vomiting. The poisonous rootstalks were used by Native Americans in a mixture of bile to poison arrowpoints (Davis 1952).

Blue-eyed Grass (*Sisyrinchium*)

General Description The four species in Idaho are perennial herbs with generally tufted, narrow leaves. The flowers have three petals and three sepals that are alike, pinkish-purple to blue in color. They are found in meadows and are often inconspicuous because of their tufted, grass-like leaves. The flowers open only in bright sunshine.

Field Notes/Uses While the uses of Idaho species is unknown, a related species *S. bellum*, was known among the Spanish-Californians as "azulea" and "villela." It was made into a tea considered to be a valuable remedy in treating fevers. It was thought that a patient could subsist for many days upon it alone (Parsons 1966).

Rush Family - JUNCACEAE

Nine genera and 400 species of *Juncus* are found in damp and wet sites of the cool temperate and subarctic regions. Two genera (*Juncus* and *Luzula*) are found in Idaho. These are grasslike annual and perennial plants with solid, rounded or flattened stems of wet and damp sites. The leaves are basal or alternate, and may be flat, folded, or round, and taper to a point. The flowers are small and have 6 undifferentiated sepals and petals (often termed tepals), 3-6 stamens, and a three parted ovary with many seeds. The family is of no direct economic importance to humans, although a few are ornamentals.

Rush (*Juncus*)

General Description and Field Notes/Uses In reviewing the scientific literature, there are no less than 34 species of *Juncus* in Idaho. In short, there quite a few of them around. Generally speaking, they are annual or perennial herbs often found in water or wet places. The flowers are in heads or panicles, and the tough, fibrous stems are inedible. However, they are useful in weaving baskets and mats.

Woodrush (*Luzula*)

General Description The species are tufted perennials with slender unbranched stems. In all, there are about 80 species of annual or perennial herbs with grass-like leaves throughout the world. The brown-green or white flowers occur in panicles. The genus name *Luzula* comes from *lux*, or light, and refers to the way the plant shines when morning dew covers its hairy leaves.

Field Notes/Uses Some species have been used medicinally as emetics and decoctions. Additionally, *L. luciola* was used as a wick in candles. The seeds of L. campestris field woodrush) are dispersed by ants because of their juicy outgrowths.

Arrowgrass Family - JUNCAGINACEAE

There are four genera and 26 species in the Arrowgrass family. Two genera occur in Idaho – *Lilaea* and *Triglochin*. There are no uses for the former genus and in general, the family is of no economic importance.

Arrowgrass (*Triglochin*)

General Description The two species in Idaho, *T. palustre* (marsh arrowgrass) and *T. maritimum* (seaside arrowgrass) are slender grasslike plants with fleshy basal leaves. Marsh arrowgrass occurs in mountain swamps and around lakes. Seaside arrowgrass is found on alkaline soils around ponds and marshes at lower elevations.

Field Notes/Uses The seeds of seaside arrowgrass can be parched and ground into flour. Roasted, they can be used as a coffee substitute. Seeds need to be parched since they contain cyanogenetic toxins that have caused death in livestock (Muenscher 1939). Parching or roasting the seeds renders them safe since the poison is volatile (Kirk 1975, Harrington 1963).

The young white leaf bases were collected around April or May from the inner leaves of the basal cluster. These leaf bases, when eaten raw at the right stage have a mild, sweet cucumber-like taste. They are generally better if cooked. In springtime, the leaf bases contain few toxic compounds, whereas the mature leaves and flowerstalks should never be eaten. The leaves contain hydrocyanic acid, a toxin that interferes with the uptake of oxygen. Symptoms include headache, heart palpations, dizziness, and convulsions. **Caution** These plants are toxic when fresh. Several references list these plants as livestock poisoners until they dry and then the cyanogenic properties evaporate, breakdown or dissipate.

Duckweed Family - LEMNACEAE

Plants in this family are small and float on slow or stagnant waters. They have small thread-like root hairs that obtain nutrients from the water. All duckweeds are used as food by wildlife and have been recorded from the stomachs of ducks. Three genera can be encountered in Idaho: *Lemna*, *Spirodela*, and *Wolffia*.

Duckweed (*Lemna*)

General Description Four species of duckweed occur in Idaho. They are small plants, often not much larger than a pinhead.

Field Notes/Uses Harrington (1967) suggests that under survival conditions, duckweed can provide copious and palatable material for salads. Strike (1994) indicates that Native Americans in California used duckweed as a diuretic and a general tonic.

Common Duckweed (*Spirodela polyrhiza*)

General Description This is the only species found in Idaho, and is frequently associated with *Lemna*. Common duckweed is a coarse species with a purplish-tinged lower side.

Field Notes/Uses As with *Lemna*, common duckweed can also provide copious and palatable material for salads. The Chinese use this duckweed to treat hypothermia, flatulence, and acute kidney infections (Meuninck 1988).

Northern Watermeal (*Wolffia borealis*)

General Description This species is native to North America. It grows in mats on the surface of calm water bodies, such as ponds. It is a very tiny plant with no leaves, stems, or roots. The green part is up to 1/16 inch long with one rounded end and one pointed end. On the flattened top of the plant is a single stamen and pistil.

Field Notes/Uses It is edible and makes a nutritious food.

Lily Family - LILIACEAE

This family contains many beautiful wildflowers. It is a large and varied family with approximately 250 genera and 4,000-6,000 species worldwide. About 75 genera are native to the United States. The family is characterized by a perianth of six parts, with a superior ovary and a 3-lobed stigma. The fruit is a capsule that splits open when ripe. The family is a source of many ornamentals, several important fibers, fermented and distilled beverages, and steroidal compounds. While some members of this family are edible, there are many poisonous species. For example, you can boil and eat the bulbs of the true lilies (*Lilium*), but some genera contain highly toxic alkaloids. Nearly two hundred alkaloids and numerous glycosides occur in the family (Tull 1987).

Onion (*Allium*)

General Description The many species of *Allium* in Idaho arise from bulbs, and all have the characteristically distinct onion odor. The odor is apparently caused by the presence of volatile sulphur compounds in all parts of the plant (causing their strong flavor and irritation to eyes). In all, there are about 300 species of onions in the world. Some other common names of *Allium* include leeks, garlic, and chives. The small flowers are clustered together in umbels. Onions are found in a variety of habitats from low elevations to the alpine zone. *Allium* is the ancient name for garlic. The derivation of this name may be from the Celtic *all*, which means "pungent."

Field Notes/Uses All *Allium* species are known to be edible. The bulbs may be eaten raw, boiled, steamed, creamed, in soup, and are especially good when used as a seasoning. Ingestion of large amounts of onions, including the cultivated ones, can cause poisoning or cause goiter, but are otherwise not known to be harmful. Regardless, eating them in moderation is the key. The plants are valuable in all seasons and can be used as greens and as flavoring. The seeds and leaves can also be eaten. Onions will keep a long time, because the skin dries and preserve the flesh inside. Wild onions do contain large amounts of some important micronutrients, more Vitamin C than an equal weight of oranges, and more than twice as much Vitamin A as an equal weight of spinach (Kindsher 1987). Additionally, onions contain a significant amount of a starch called inulin, which is not easily digested by humans.

Medicinally, onions have a number of uses. Soldiers during World War I took advantage of their natural antiseptic properties by applying *Allium* juice to wounds to prevent infection. The juice of wild onions can be boiled down until it is thick and used as a treatment for colds and throat irritations (Willard 1992). The juice was also used as an insect repellent when rubbed over the body. The onion smell apparently has some beneficial effects on the circulatory, digestive, and respiratory systems.

Warning Wild onions should not be confused with the so-called "poison onions" or death camas (*Zigadenus*). These are bulb-bearing plants with grasslike leaves, also in the Lily family. They have upright, more elongated (not umbrella-like) clusters of white or cream flowers. They contain highly toxic alkaloids and all plant parts, including the bulbs, can be fatal if ingested in any quantity. They also lack the characteristic strong odor of onions.

Garden Asparagus (*Asparagus officinalis*)

General Description This is a perennial plant with stems up to 5 feet tall. The alternate, membranous, scale-like leaves subtend several, short, filiform branches that resemble needles of a conifer. The drooping flowers occur in the axils of the leaves. The 3 petals and 3 sepals are similar and greenish-white in color. The fruit is a few-seeded, red berry.

Field Notes/Uses Having been grown in cultivation for many years, asparagus has escaped and can be found almost anywhere at the lower elevations. Specifically, the plant thrives on neglect and can be found growing along roadsides, vacant lots, railroads, and disturbed sites. The small flowers are produced in the early summer. The shoots purchased and eaten are simply the young stems of the plant which appear in the early spring.

Asparagus is a perennial plant native to Europe and East Asia which has been cultivated since the time of the ancient Greeks, becoming widely spread as a result. It is grown from seeds or rhizomes for its young succulent shoots which are eaten as a vegetable. Useful variants are maintained by clonal propagation, since seeded crops are highly variable.

The virtues of asparagus are well known as a diuretic and laxative; and for those of sedentary habits who suffer from symptoms of gravel, it has been found very beneficial, as well as in cases of dropsy. The fresh expressed juice is taken medicinally in tablespoonful doses. The smell in one's urine after eating asparagus is caused by excretion of the substance methyl mercaptan.

Mariposa Lily, Sego Lily (*Calochortus*)

General Description Seven species of *Calochortus* are known to occur in Idaho. They are characterized as perennials from bulbs,

Digging Sticks

Roots and bulbs of many plants are best collected with the aid of a digging stick. A digging stick is about 2-3 feet long, 1-2 inches thick, beveled at one end and fire hardened. To use, thrust the stick into the ground beside the plant and pry upward while pulling on the plant from above.

To fire harden your digging stick, simply hold the point a few inches above a bed of hot coals and slowly turn it as you would a skewer. Take care not to char the wood, but let it turn to a light brown color.

with tulip-like flowers that are few and showy. Mariposa is the Spanish name for butterfly. These species can be found in dry open places from low to mid elevations.

Field Notes/Uses While the entire *Calochortus* plant is edible and can be used as potherbs, the highly nutritious bulbs are usually sought. They are smaller than walnuts and may be eaten raw, boiled or roasted in hot ashes in pits, or steamed before eating. The bulbs are dug in the early spring, usually before flowering. The bulbs can also be threaded on a string and dried with or without cooking first. They can also be dried and ground into flour or cooked and mashed into cakes for preservation. *Calochortus* bulbs were eaten by many Native Americans and were considered an important food source. When numerous bulbs were collected, they were usually pit-cooked. The flower buds can be eaten and have a sweet taste. Seeds are also edible.

However, care should be taken not to over harvest these plants. This is particularly true in Utah where *C. nuttallii* (sego lily) is the State Flower.

Note These plants are becoming increasingly rare in some areas, primarily due to habitat destruction and overgrazing. Harvesting the corms destroys the plants. Because of the plants rarity and beauty, their use today is not recommended.

Camas (*Camassia*)

General Description Ttwo species of *Camassia* occur in Idaho, but it is *C. quamash* (common camas) that was frequently utilized by Native Americans. Arising from a bulb, it has bright blue to violet, 6-parted flowers in a showy spike-like raceme. The leaves are basal and grass-like. The plant can be found in meadows, marshes, grassy slopes, and fields from low to mid elevations. The plant's common name is derived from the Nootka Indian word, *chamas*, which means sweet.

Field Notes/Uses Camas was perhaps the most important food of Native Americans in western North America. As Gunther (1973) writes;

> "...except for choice varieties of dried salmon there was no article of food that was more widely traded in western Washington then camas."

Camas was an important staple food to many aboriginal peoples, and the bulbs were eaten wherever available. In fact, some tribes managed or "owned" areas where the plant grew. They would intentionally set fires to ensure a bountiful harvest. As testimony to the species' former abundance, Meriwether Lewis in his journal entry of June 12, 1806, stated:

"The quawmash is now in blume and from the colour of its bloom at a short distance it resembles lakes of fine clear water, so complete in this deseption that on first sight I could have sworn it was water."

Camas bulbs were dug out of the ground from late July through September with digging sticks. The black outer covering of the bulb is removed, and the white bulbs are then steamed or cooked in pits for 24 hours or more. When cooked this way, the bulbs turn dark brown, become quite moist and soft, and sweet. Cooking is required because the plant contains a carbohydrate called inulin. Inulin is not very digestible or very palatable in its "raw" form. Cooking is necessary to chemically breakdown the inulin into its component fructose sugar. Common in fruits and honey, fructose is both easily digested and sweet tasting. After cooking, the camas bulbs can be mashed and dried into cakes for storage. They are considered to be more nutritious than potatoes. The bulbs can also be boiled down into syrup. Too much camas, however, is both an emetic and purgative (Willard 1992).

Caution When collecting bulbs, be aware that the poisonous death camas (*Zigadenas*) may be in the area too.

Bride's Bonnet (*Clintonia uniflora*)

General Description This is a perennial from a creeping rhizome with basal leaves. The flowers are white, large, and cup-shaped, and the berries are blue. The plant is found in various forest habitats from low to subalpine elevations.

Field Notes/Uses The berries of bride's bonnet are considered to be unpalatable (Pojar and MacKinnon 1993). Elias and Dykeman (1982) suggest gathering the leaves before they fully unfurl, then chop, boil, and use them in salad. A related species in the eastern part of the United States, *C. borealis* (bluebeard lily), was also used as a potherb and has a cucumber-like taste. A poultice of roasted or mashed leaves of *C. uniflora* was used for sore eyes and cuts (Strike 1994).

Avalanche Lily, Glacier Lily (*Erythronium grandiflorum*)

General Description This is the only species in Idaho. The bright yellow flowers hang from a slender stalk. The two flat leaves are sheathing at the base. Glacier lily can be found up to and above timberline, blooming at the edge of melting snowbanks. Linneaus derived the genus name from the Greek word meaning red, referring to a red-flowered European species.

Field Notes/Uses These beautiful flowers are seldom abundant, so glacier lily should only be considered in extreme emergencies. Harvesting destroys the plant. The young plants can be boiled as a potherb. The seed pods can be eaten raw or cooked. Eating the seed pod will not destroy the plant as long as you spread the seeds around first. Corms may be eaten raw, but are better if boiled for at least 20 minutes. They can also be dried and stored for future use. The corms contain inulin, which is inedible raw. Normally they would be pit cooked for an extended period of time.

Medicinally, the crushed bulbs were used by Wailaki to poultice boils (Strike 1994). An infusion of leaves has been shown to be active against a wide spectrum of bacteria (Tilford 1997).

Missionbels, Fritillary (*Fritillaria*)

General Description Three species, *F. pudica* (yellow missionbells), *F. atropurpurea* (spotted missionbells), and *F. lanceolata* (spotted missionbells),

occur in Idaho. They are glabrous, perennial herbs from bulbs with numerous white bulblets around the base. The flowers are usually nodding with similar petals and sepals (tepals). The genus name comes from the Latin *fritillus*, meaning "dice box," in reference to the short, broad capsule characteristic of the genus. Fritillary can be found in open areas, forest, meadow, and grassland habitats at low and middle elevations.

Field Notes/Uses The bulb has been a staple for native peoples since prehistoric times. Bulbs of all species are edible raw or cooked, but are relatively rare and should be considered only in an emergency. The fruit pods of yellow missionbells can be eaten as a potherb.

Common Starlily (*Leucocrinum montanum*)

General Description Common starlily is a stemless perennial from short, fleshy roots. Each plant has about a dozen grasslike leaves about six inches long that are longer than the flowers, and are surrounded at the base by papery white sheaths. The fragrant white flowers are up to one and a half inches wide, and include a slender tube nearly four inches long. At the bottom of the tube are the ovaries and seed capsules which mature below ground. Seeds are black.

Field Notes/Uses The generic name was compounded from the Greek *leuco*, "white", and *krinon*, "lily." The specific epithet *montanum* means "of mountains" in Latin. Common starlily was described for science in the early

1800's by the famous English botanist-naturalist Thomas Nuttall (1786-1859). Medicinally, a poultice of pulverized roots was applied to sores or swellings by the Paiute. The plant was also used as food by the Crow.

Lily (*Lilium columbianum*)

General Description The true lilies have large showy flowers that are usually spotted. The flowers may be funnel-shaped, or the tepals maybe curved backward. The color varies from yellow to orange, with one to several flowers appearing at the top of a single stalk up to 3 feet tall. Leaves are alternate or whorled. The roots are covered with scales, an important identifying characteristic.

Field Notes/Uses The bulbs of all members of this genus are edible. They can be eaten raw or steamed, and have a bitter or peppery taste. Some suggest that the plants are better used as a flavoring agent and that the bulbs taste better after flowering. One way to locate bulbs in the fall is to flag the plant(s) during the summer. Because of their relative rarity and beauty, they should not be harvested except in an emergency.

Common Alplily (*Lloydia serotina*)

General Description The alplily is a diminutive plant that is often over-looked. It has one to several stems arising 4 to 8 inches tall from bulbs on short, thick rootstocks. The basal leaves are needle-like. The old, withered leaf bases remain at ground level, simulating a bulb. The 3-4 short stem leaves are wider and the one (usual) to two flowers are found atop the scape and are usually held erect. The flowers are whitish with green or purplish veins. The 6 tepals are oblong or oblanceolate in shape and are usually about $\frac{1}{2}$ inch long.

Field Notes/Uses This dainty lily grows high in alpine meadows, often abundantly, and might at first be taken for spring beauty (*Claytonia*). Notice also that its long thin leaves might be taken for wild onion (*Allium*) leaves. *Lloydia* is for Edward Lloyd (1660-1709), Curator of the Museum of Oxford University and discoverer of this plant. The species name *serotina* is Latin for "late."

Mayflower, False Solomon's-seal (*Maianthemum*)

General Description Three species occur in Idaho: *M. dilatatum* (false lily of the valley), *M. racemosum* (feathery false lily of the valley), and *M. stellatum* (starry false lily of the valley). The latter two were previously in the genus *Smilacina* and they are perennial herbs with extensive rootstalks and unbranched stems that are scaly below and leafy above. The leaves are alternate and sessile. They are more widespread in the state and can be found in thickets and open or shady forests up to the subalpine zone. M. *dilatatum* is more localized in the northern part of the state. It has two or three leaves are heart-shaped and the flowers are small, white in dense terminal clusters. The pea-sized berries are initially hard and green with brown mottling, then becoming soft and red as they mature. The plant is found in moist woods and clearings in the Panhandle part of the state.

Field Notes/Uses The new "folded" leaves of M. *dilatatum* were boiled and eaten as greens (Norton 1981). The berries are edible, but not highly regarded. The other two species have edible berries that are not especially palatable. If eaten in quantity they can act as a laxative. Cooking the berries removes much of the purgative elements making them a bit more palatable. They are also high in Vitamin C. The young shoots and leaves can be used like asparagus or eaten as a potherb.

M. *racemosum* have starchy rootstocks that may be eaten. However, the rootstocks must be soaked overnight in lye. The Ojibwa Indians of Ontario, Canada, used the white ashes from their fire pits instead of lye (Willard 1992), which supposedly removed the bitterness. The roots are then boiled and rinsed several times to remove the lye. A tea made from the roots was used for headaches (Moore 1979).

A tea from M. *stellatum* was used by some Native Americans as a contraceptive (Strike 1994). The powdered roots were used on wounds to stop bleeding, and a root decoction was used internally as a tonic or externally as an antiseptic wash for infected sores or wounds. The mashed root of M. stellatum was thrown into a stream as a fish stupifier, making the fish easier to catch (Strike 1994). A decoction of M. *racemosum* was used as a contraceptive, to regulate menstrual disorders, to relieve kidney problems, to heal wounds and as a heart tonic.

Fairy-bells (*Prosartes*)

General Description The two species in Idaho, P. *hookeri* (drops of gold) and P. *trachycarpum* (roughfruit fairybells), are characterized as rhizomatous perennials with branched, leafy stems. The sessile leaves are prominently veined and slightly rotated where they meet the stem. The flowers hang from the tips of the branches and are tubular to bell-shaped. They can be found in moist to somewhat dry forest areas from foothills to the subalpine zone. Both species were once classified under the genus Disporum.

Field Notes/Uses The velvet-skinned yellow or orange berries of *P. trachycarpum* can be eaten raw (Kirk 1975, Weedon 1996). The fruit of *P. hookeri* was used to relieve kidney ailments (Strike 1994).

Twisted-stalk (*Streptopus*)

General Description There are two species of twisted-stalk in Idaho, *S. amplexifolius* (claspleaf twisted-stalk) and *S. streptopoides* (northern twisted-stalk). They are characterized as herbs with creeping rootstocks. The sessile or clasping leaves are alternate, elliptical to ovate in shape, and the flowers are yellowish-green. Common in moist soil and along streams and thickets in the montane and lower subalpine zone, they are often associated with genera such as *Smilacina* (false solomon's-seal) and *Actaea* (baneberry).

Field Notes/Uses In terms of edibility, these plants have escaped mention in many guides but are indeed safe. The new spring shoots and clasping young leaves can be eaten raw or added to salads and taste somewhat like cucumbers. The berries, often referred to as watermelon berries are somewhat laxative if eaten in excess, but may be eaten raw or cooked in soups and stews. They are sometimes referred to as "scooter berries," because if you eat too much you can find yourself "scooting" off to the bathroom. The species are easy to grow in wild gardens. The stems were used in poultices for cuts (Pojar and MacKinnon 1994).

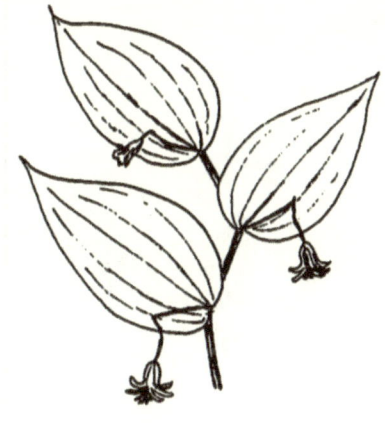

Warning Anyone wishing to use the young shoots of twisted-stalk should be very careful to identify it correctly. At the shoot stage, these plants resemble the highly toxic *Veratrum* (false hellebore).

Western False Asphodel (*Triantha occidentalis*)

General Description This unique plant can be found near bogs and marshes. It has numerous, narrow, lily-like leaves growing in a basal clump, or slightly up the stem. The erect, green stem is sticky. The dense terminal clusters of white flowers have dark anthers which extend beyond the petals. This species was previously known as *Tofieldia glutinosa*.

Field Notes/Uses The genus name honors Thomas Tofield (1730-79), a British botanist. The genus contains about 18 species of rhizomatous herbs that have narrow, tufted, basal leaves and racemes of small star-shaped flowers. Some species are cultivated ornamentally.

Wakerobin (*Trillium*)

General Description Two species of wakerobin may be found in Idaho, *Trillium ovatum* (Pacific trillium) and *T. petiolatum* (Idaho trillium). They are low, glabrous herbs from short, fleshy rootstocks. The stems bear a whorl of three sessile, broadly egg-shaped leaves, and the single stalked flower is borne above them. The sepals are green, and narrower and shorter than the snow-white petals. *Trillium* species are commonly found in moist, rich soils of forests.

Field Notes/Uses The stem and leaves may be boiled and eaten as greens. The leaves should be collected before they fully unfold because when the flowers appear, the leaves become bitter. The berries and roots are inedible, possibly possessing emetic properties (Peterson 1977, Fernald and Kinsey 1958). Because of their relative rarity and beauty, refrain from using them except in an emergency situation.

The juice from Pacific trillium and a poultice made from the leaves and roots were used for boils (Strike 1994). Decoctions of the plant were used to treat nearly every illness and to prevent deep sleep (Strike 1994).

Triteleia (*Triteleia*)

General Description The two species of *Triteleia* (previously *Brodiaea*) found in Idaho are erect herbs arising from corms. The leaves are linear and the flowers are in umbels. They can be found in dry to moist soil, in rocky areas, and meadows from low to mid elevations.

Field Notes/Uses The corms of most *Triteleia* species are edible raw, but are somewhat muscilagenous. It is better if they are boiled for a few minutes or roasted. They can also be mashed and dried for future use in stews. Since the corms grow deep, it is usually easiest to harvest them with digging sticks. The tender seed pods of *T. grandiflora*) (largeflower triteleia) can be eaten as greens (Craighead et al. 1963).

The crushed corms were used as a paste which was smoothed over the sinew backing on bows. The paste was also used to bind paint pigments to hunting bows. *Triteleia* corms and flowers were used as soap and shampoo (Strike 1994).

Corn Lily (*Veratrum*)

General Description The two species that occur in Idaho, *V. californicum* (California false hellebore) and *V. viride* (American false hellebore), are characterized as large, coarse, leafy-stemmed herbs with thick rootstocks. The leaves are broad, clasping, and strongly veined. The numerous green to dull white flowers are arranged in a large panicled inflorescence. Both species are found in wet meadows and forest openings from the middle elevations into the alpine zone.

Field Notes/Uses These plants are very poisonous if ingested and have an inconsistent mixture of several powerful alkaloids (Kingsbury 1964, Muenscher 1940). Some of the symptoms include depressed heart action, salivation, headache, burning sensation in the mouth, slowing of respiration, and death from asphyxia. These violent symptoms of poisoning may occur within 10 minutes. Willard (1992) and others suggest avoiding any use of the plant that involves ingestion. In some cases, just handling *Veratrum* can cause severe itchiness and irritation. The water in which the roots of both California false hellebore and American false hellebore were boiled was considered effective in killing head lice. The powdered roots have also been used as an insecticide. Even nectar in the flowers is poisonous to insects and can cause serious losses among honeybees.

Common Beargrass (*Xerophyllum tenax*)

General Description This is a stout perennial up to five feet tall. The stems arise from large clumps of bluish-green, wiry, saw-edged, grasslike leaves that are up to two feet long. The cream-colored flowers are borne in a hemispheric terminal inflorescence. Common beargrass is found in mid to lower alpine open slopes and forests. The thick, shallow rhizome can remain vegetative for many years, then flowers and dies.

Field Notes/Uses The fibrous roots of common beargrass can be eaten after roasting or boiling. Although the sharp leaves are not very pleasant to handle, when dried and bleached they can be used for weaving baskets and

clothing. The baskets are particularly pliable and durable. Lather from the roots was used to bathe sores (Strike 1994).

Death Camas (*Zigadenus*)

General Description The three species of *Zigadenus* in Idaho are glabrous perennials with bulbs and grasslike leaves. The cream-colored to greenish white flowers are stalked and subtended by narrow bracts in a elongated inflorescence. Death camas occurs in grasslands, meadows, and forest openings into the alpine zone.

Field Notes/Uses Death camas is very poisonous plants if ingested. The alkaloids, primarily concentrated in the bulbs, can cause muscular weakness, slow heartbeat, subnormal temperature, stomach upset with pain, vomiting, and diarrhea, and excessive watering of the mouth. Death camas should not be confused with the edible camas (*Camassia*), which formed a staple food for aboriginal peoples in the Northwest. It is also difficult to distinguish death camas from other edible plants, including wild onion (*Allium*), sego lilies (*Calochortus*), fritillaries (*Fritillaria*), and triteleia (*Triteleia*) prior to flowering.

Crushed death camas bulbs were used by some Native Americans as poultices for boils, bruises, strains, rheumatism, and in some cases rattlesnake bites (Balls 1962).

Orchid Family - ORCHIDACEAE

This is one of the largest families in the world with the greatest concentration of species found in the tropics. There are over 15,000 species in several hundred genera. In Idaho, there are eleven genera and 29 species of orchids. The flowers are irregular in shape with three sepals and three petals. One of the petals forms a lip, sac or pouch on the lower side of the flower. The flower structure is highly specialized for insect pollination. The family is an outstanding source of ornamentals, and although a few orchid species have utilitarian uses, many of the species are considered rare, and should only be considered for use in emergency situations.

Fairy-slipper Orchid (*Calypso bulbosa*)

General Description This is a perennial arising from a corm. The flowers are rose-purple, large, showy, and solitary, with a large and slipper-like lower lip. It is found in forests at low to middle elevations. The genus name is Greek for Kalypso, the sea nymph in Homer's Odyssey, for her beauty and secretive behavior.

Field Notes/Uses The small corm of the fairy-slipper orchid can be eaten raw, roasted or boiled. However, because the plant is rather rare, it should only be considered in a dire emergency.

Phantom Orchid (*Cephalanthera austiniae*)

General Description This is a very rare orchid and is the only member of the genus *Cephalanthera* that is in North America. The plant is terrestrial, mycoheterotrophic, leafless and completely white. Mycoheterotrophic plants are nonphotosynthetic parasites using fungal intermediaries to withdraw nutrients from other plants. The plant can stay dormant for years before sending out as stem. It may bloom once and then go into remission for up to 17 years.

Field Notes/Uses The flowers are erect, white with a small yellow spot on the lip; very similar to *Epipactis* but with a longer and more slender column. The habitat is in the rich soils of dark coniferous or deciduous forests along mountain ranges. They only grow in forests with little to no undergrowth and limestone deposits.

Threats to the species survival are also increasing through habitat loss. High levels of development and logging in key orchid habitats continue to place this species in peril, as well as mountain bike activity and native plant gathering.

Coralroot (*Corallorrhiza*)

General Description Five species of coralroot can be found in Idaho. They are typically yellowish to brownish-red perennials with coral-like rhizomes. *Corallorrhiza* means "coral root." The rhizomes are associated with fungi that aid in the uptake of nutrients.

Field Notes/Uses Consumption of the toxic rhizome can cause hyperthermia and profuse perspiration (Weedon 1996). The rhizome of *C. maculata* (summer coralroot) has been used as a diaphoretic, febrifuge, and a sedative. The dried stems were used by the Paiute and Shoshone Indians of

Nevada to make a tea to build up blood in pneumonia patients (Coffey 1993, Foster and Duke 1990). Strike (1994) indicates that the plant was used to reduce fevers or as a sedative.

Lady's Slipper (*Cypripedium*)

General Description Three species of lady's slippers are known to occur in Idaho. They are easily recognized by their large inflated lip, and they are among the rarest plants. The flower attracts insects into the pouch from which there is only one exit. The insect must pass over the stigma where pollen from previously visited flowers is brushed off, and then go under one of the anthers where new pollen is picked up. This highly advanced pollination strategy reduces the chance of self-fertilization.

Field Notes/Uses The rhizome of *C. montanum* (mountain lady's slipper) is said to be toxic, but has been used medicinally as a stimulant and antispasmodic (Weedon 1996). It has been used with good results in reflex functional disorders or chorea, hysteria, nervous headache, insomnia, low fevers, nervous unrest, hypochondria, and nervous depression accompanying stomach disorders Willard (1992). The plants were also used by Native Americans as tranquilizers. A related species, *Cypripedium calceolus* var. *pubescens* (= *C. pubescens*) was widely used as a sedative for nervous headaches, hysteria, insomnia, and nervous irritability (Foster and Duke 1990). Many species appear to cause dermatitis.

Giant Helleborine (*Epipactis gigantea*)

General Description This is a terrestrial plant arising from a rhizome. The lanceolate leaves clasp the lower stem. The flowers are pink to rose and the sepals are greenish to rose-color with dull purple to red nerves. It is found in marshes, along stream and rivers, and in meadows. The name Epipactis is derived from a classical name used by Theophrastus (circa 350 B.C.) for a plant used to curdle milk, or it may refer to the Greek "*epi*" for on or upon and "*pactos*" for solid or firm. The specific epithet refers to the Latin for gigantic in referencing the large size of the plant.

Field Notes/Uses Vegetative plants may be confused with some members of the orchid genus *Platanthera*, or more likely with *Maianthemum stellatum* (Liliaceae), species that can co-occur with giant helleborine. The prominently clasping leaf bases and taller habit of giant helleborine distinguishes it from Maianthemum, and its generally more numerous and larger leaves and taller habit from *Platanthera*. A decoction from the root of giant helleborine was used by the Native Americans in California for illnesses which were manifested by overall soreness or paralysis (Strike 1994).

Western Rattlesnake Plantain (*Goodyera oblongifolia*)

General Description This is a perennial herb with glandular-hairy stems. The persistent basal leaves are dark green and have winged petioles and broadly lance-shaped blades that are mottled with white along the midvein. The flowers are greenish-white. It is found in open or deep forests up to the subalpine elevations. A second species, *G. repens* (lesser rattlesnake plantain), may also occur in Idaho

Field Notes/Uses The herbage of western rattlesnake plantain is slightly toxic if eaten (Weedon 1996). The plant was also known by some aboriginal peoples as a medicine for childbirth, and as a poultice for cuts and sores in which the leaves were split open, and the moist inner part of the plant was placed over the wound (Pojar and MacKinnon 1994).

Twayblade (*Listera*)

General Description These are small, slender, often rhizomatous perennials with 2 broad, prominently nerved leaves opposite each other in the middle of the stem. Small green or yellowish flowers occur in a short, narrow, usually glandular-hairy inflorescence. The 3 sepals and 2 upper petals are similar and spreading or reflexed. The prominent lip petal projects down or outward. These are small orchids.

Field Notes/Uses The intricate pollination mechanisms of Listera species fascinated Charles Darwin, who studied them intensively. The central furrow holds nectar with leads insects towards the stigma. The pollen is blown out explosively within a drop of viscous fluid that glues the pollinia to unsuspecting insects (or to your finger if you touch the top of the column). The pollen will then be deposited on the next twayblade they visit. The genus *Listera* is named in honor of Dr. Martin Lister, an English naturalist who lived from 1638-1711.

Rein Orchid (*Piperia*)

General Description These are perennials with erect, unbranched stems. The 2-5 leaves are basal, but the cauline leaves are much reduced in size.

In general, these orchids are found in wet habitats. The genus name honors Charles Vancouver Piper (1867-1926), an agronomist with the U.S. Department of Agriculture and an expert on Pacific Northwest flora. Three species occur in Idaho: *P. elegans* (elegant piperia), *P. elongata* (denseflower rein orchid), *and P. unalascensis* (slender-spire orchid).

Field Notes/Uses The bulbs of *P. unalascensis* were baked and eaten like baked potatoes by the Pomo and Kashaya.

Rein Orchid (*Platanthera*)

General Description The six species in Idaho are perennials, often with fleshy or tuberous roots. The small white to yellowish-green flowers are in spike-like racemes. At the base of the lip is a spur. The species here were once classified under the genus *Habenaria; which is* Latin for "reins" or "narrow strap" which refers to the lip of some species. Rein orchids can be found in forest understories, or in wet areas and meadows.

Field Notes/Uses The tuber-like roots of many species may be eaten raw or cooked (Coffey 1993, Kirk 1975, Weedon 1996). However, Pojar and MacKinnon (1994) recommend a cautious approach until the poisonous nature of the plants is clarified. Some Native Americans used extracts from these plants as poison to sprinkle on baits for coyote and grizzly bear.

Hooded Ladies'-tresses (*Spiranthes romanzoffiana*)

General Description This is one of three species found in Idaho. It is a small herb with fleshy roots. The white flowers are in a dense spirally twisted spike, with the sepals and lateral petals appearing to be fused and forming a hood around the column. Hooded ladies'- tresses can be found in dry to moist areas, meadows, lakeshores, bogs, and marshes, up to timberline. The species was named after Nikolei Rumliantzev, Count Romanzoff, a Russian patron of science.

Field Notes/Uses According to Weedon (1996), *S. romanzoffiana* has strong diuretic properties, making it undesirable for eating. Additionally, Foster and Duke (1990) indicate that Native Americans used a plant tea of *S. cernua* (nodding ladies'-tresses) of eastern United States as a diuretic for urinary disorders, venereal disease, and as a wash to strengthen infants. They also state that other North American, European, and South American species have also been used as diuretics and aphrodisiacs.

Grass Family - POACEAE

With 600 genera and 10,000 species of grasses worldwide, the Grass family is the most common family of flowering plants found in practically all habitats and on all continents. There are over 180 genera and nearly 1,000 species in the United States. Grasses have round, hollow stems with linear sheathing leaves. Because grasses are wind pollinated, they have no showy flowers to attract insects. The flowers have been reduced to scaly bracts that enclose the male and female parts. The grains form within the papery bracts after pollination. The grass family contains many of the most economically important plants in the world, including all the important cereal crops and forage grasses essential to raising domesticated livestock. Grasses can be found from alpine meadows sea level.

All grasses in Idaho have edible grains that are generally small and tedious to collect. The small seeds are tightly enclosed in scales which are hard to remove. The larger grains of some grass species were a staple among aboriginal peoples. The grains were harvested with a beater and ground for mush and flour. The grains are rich in protein and can be eaten raw, but are better if roasted, ground into flour, or boiled into mush. They may also be boiled in the same way as rice, and added to soups or stews. The reported toxicity of some grass species may be that of a fungus (i.e, *Claviceps purpurea*) associated with the grasses. Any inflorescences containing black grains should be discarded since they may have a harmful fungus infection.

Brown (1983) suggests that all bladed grasses are edible and are rich in vitamins and minerals. Animals often consume grasses to get nutrients they can't get elsewhere. The young shoots are edible raw, and are not as fibrous, and therefore, easier to digest than mature leaves. The green or dried leaves can be steeped to make a tea. Following are some genera of grasses used by Native Americans.

Note Grasses that are infected with *Claviceps purpurea* develop purple sclerotia (ergots) in place of the healthy grain. The sclerotia contain a number of toxic alkaloids and if eaten, can cause severe illness and sometimes death. One effect of the toxins is

Wild Flours

Pinole -- was made using small grains which were parched by tossing in a basket with glowing coals or hot pebbles, keeping the grains in constant motion. Grains were then pulverized and eaten. Sometimes the pulverized grains were pressed into cakes held together by the grains' natural oil, with no other liquids added. Grains from several different species of plants were often mixed together to enhance the flavor constriction of the blood vessels, whereby the impaired circulation may result in gangrene or loss of limbs. Another effect is on the nervous system, which results in convulsions and hallucinations (Webster 1980). The sclerotia of *C. pupurea* are used medicinally to hasten uterine contractions during childbirth. The ergot of commerce is produced by cultivating the fungus on rye (*Secale cereale*) and other plants. Attempts are being made to extract the medically important alkaloids from pure cultures of the fungus (Webster 1980).

Holy Grass, Sweet Grass (*Hierochloe odorata*)

General Description This is a sweet scented perennial with purplish-based culms. The yellowish brown spikelets are three flowered, and are borne on spreading branches in an open, pyramidal inflorescence. Holy rass can be found in moist or wet meadows from lower elevations to the lower alpine zone.

Field Notes/Uses Holy grass was used by Native Americans as incense in religious ceremonies. The plant contains a glycoside which when dried produces coumarin, a sweet smelling crystalline compound important in perfumes. The leaves can also be added to vodka as flavoring. Holy grass is in demand for sachet and basket weaving because it holds its fragrance for years. The oil from the plant was used to flavor candy, soft drinks, and tobacco. The stems were soaked in water and used to treat chapping and wind burn, and as a tea for coughs and sore throats (Willard 1992, Foster and Duke 1990).

Muhlenbergia (*Muhlenbergia*)

General Description The six species of *Muhlenbergia* in Idaho are annuals or rhizomatous perennials with narrow bladed leaves. The spikelets are one-flowered in an open inflorescence. *Muhlenbergia* are usually found in wet areas from the mid to high elevations.

Field Notes/Uses All species were widely used in basket making. The stalk of the plant was used as the horizontal or fundamental elements around which the coils were wrapped.

Indian Ricegrass (*Oryzopsis hymenoides*)

General Description A perennial bunchgrass from 8 to 20 inches tall, Indian ricegrass is found in dry rocky or sandy ground, grasslands, and valleys from low to mid elevations.

Field Notes/Uses The relatively large grains of this species have been used by Native Americans as food for centuries. The grains were collected with the stems and held over a fire to singe off the fine white hairs. They can also be collected in a pan or basket with hot coals or rocks, and shaken to burn off the hairs. The grains were then ground into flour and used as mush, to thicken soup, or made into cakes.

Common Reed (*Phragmites communis*)

General Description Common reed is native to every continent except Antarctica. It often forms dense thickets along streams, ditches, and marshes at low elevations.

Field Notes/Uses The record of human uses for this species is extensive. The reeds have been used for roof insulation since before the time of Christ. Because it is light, it is best known for its use in arrow shafts. The roots can be eaten raw or cooked.

Meal can be obtained from pulverized stems. In the fall, the leaves and stems may become encrusted with grayish exudate. This exudate, actually honeydew (excreta of whitefly and aphids), was obtained from stalks. Stalks were cut and flayed to remove honeydew crystals, which were winnowed and cooked into stiff dough. Dough was formed into cakes, sun dried, and stored. Split culms provided fiber. Common reed was used to make flutes and other musical instruments in addition to carrying nets and cordage. The honeydew was given to pneumonia patients to loosen phlegm and soothe pain in lungs.

Pondweed Family - POTAMOGETONACEAE

Of the two genera and 100 species worldwide, *Potamogeton* is native to the United States. The family is of no direct economic importance to humans, but provides a valuable source of food for ducks and other wildlife.

Pondweed (*Potamogeton*)

General Description Pondweeds are perennial aquatic plants with extensive, slender rhizomes and simple branched stems that often root at the nodes. The leaves have stipules that clasp the stem. The lower leaves are alternate and submerged, and the upper often floating leaves are wider and opposite. The small, greenish flowers are clustered in a spike that arises from the upper leaf axils. Most Pondweed species can be identified without flowers and fruits, but the floating or submersed leaves may be necessary. Hybridization among species is fairly common.

Field Notes/Uses Many species of *Potamogeton* occur in Idaho. Probably all pondweeds have starchy, edible rhizomes, but species with larger rootstocks are preferred for gathering. *Potamogeton diversifolius* (waterthread pondweed) was an important source of strong fibers which were rolled into cordage to make carrying nets, rabbit-trap nets, and other items (Strike 1994).

Bur-reed Family - SPARGANIACEAE

One genus with about 20 species occurs in this family. Bur-reeds are found primarily in the cooler regions of the North Temperate zone, Australia, and New Zealand. Several species are native to the United States. The plants arise from creeping rootstalks and the roots are fibrous. The species have no direct economic importance to humans. Recent studies now lump this family into the Cattail family (Typhaceae).

Bur-reed (*Sparganium*)

General Description The five species of bur-reed in Idaho are aquatic perennials with unbranched, erect or floating stems. The leaves are linear and sheath the stem. The flowers are borne in dense, round clusters. Bur-reeds can be found in shallow waters of marshes, ponds, and slow moving steams.

Field Notes/Uses The bulbous bases of the stems and tubers of *S. eurycarpum* (broadfruit bur-reed), *S. simplex* (simplestem bur-reed) and *S. angustifolium* (narrowleaf burreed) can be used as food in much the same way as cattails (*Typha*) and bulrushes (*Scirpus*): dried and pounded into flour.

Cattail Family - TYPHACEAE

There is one genus (*Typha*) with about 15 species worldwide in this family. Members are marsh or aquatic perennials with creeping rootstalks. Two kinds of flowers are borne in crowded spikes, with the staminate (male) flowers above and the pistillate (female) flowers below. The family is of little economic importance.

Cattail (*Typha*)

General Description The two species, *T. angustifolia* (narrowleaf cattail) and *T. latifolia* (broadleaf cattail), found in Idaho are also found over much of North America. *Typha* is Greek for "cattail." It is often referred to as "Cossack asparagus." Cattails reproduce rapidly in marshy areas that are often unsuited for agriculture.

Field Notes/Uses In his book Mountain Man, Vardis Fisher (1965) illustrates how Indians utilized cattails:

"And what be that?' asked Sam, staring at the mold. He knew that Indians ate just about everything in the plant world, except such poisons as toadstools, larkspurs, and water parsnips. It was a marvel what they did with the common cattail - from the spikes to the root, they ate most of it. The spikes they boiled in salt water, if they had salt; of the pollen they made flour; of the stalk's core they made kind of a pudding; and the bulb sprouts on the ends of the roots they peeled and simmered."

Virtually every part of the plant has a use, from food to fiber. Euell Gibbons considered the cattail the "supermarket of the swamps." Although both cattail species have edible rhizomes, the rhizomes should never be raw since they may cause vomiting. The rhizomes should be boiled or roasted, or dried and then ground into meal or flour. Another way to obtain the starch from the rhizomes is to follow a technique described by Euell Gibbons (1962):

"...after scrubbing the root [i.e., rhizome] and peeling off the spongy outer layer surrounding the white stiff core, cut the core into small sections, and place the pieces in a bowl of cold water. Work the core with your hands, separating the fibers and scraping out the starch. Slosh the fiber around in the water until you have removed all the starch. Pour off the water through a course sieve to extract the fibers. Allow the water to settle for a little while the starch settles to the bottom of the container. Then carefully pour off the water, leaving the starch in the bowl. For a cleaner starch, pour in some more water and let it settle again. Then pour off the water. After this, you can use the starch almost immediately to make pancakes, breads, and biscuits."

Harrington (1967) reports that one acre of cattails in a marsh can yield over three tons of nutritious flour. However, to make it profitable, extraction techniques would need to be refined for commercial exploitation.

When pulling up the rhizome, you may notice newly emerging buds. These can be scrubbed, peeled, and eaten raw or boiled. The swollen joint between bud and rhizome is also starchy. Peel it, then roast or boil for a potato-like vegetable. Like the rhizomes, this part should not be eaten raw. The young green shoots can be peeled of their green outer layer and eaten raw or cooked. It is always good to boil them in a couple of changes of water if there is any bitterness. The peeled core can also be sliced and added to salads.

While the flower spikes are still green, remove the papery sheath and boil the cluster for a few minutes. The flower spikes can then be eaten like "corn on the cob", although the core of the cluster is inedible. Cattail pollen is high in protein, and can be used in flour for breads or eaten raw. However, if you are allergic to pollen, it should be avoided. The seeds from the female portion of the flower spike can be pulverized to make nutritious, protein rich flour. Seeds can be extracted from the fluff by parching them.

Useful fibers can be derived from cattails. Fibers in stems can be loosened by soaking plant material in water for several days. The silky fluff on the seeds is buoyant and water repellent and makes a good insulator, especially in boots. The silk can be used for stuffing items from pillows to down vests. It can also be used for tinder. The fuzz will explode into flame with a spark from a flint and steel set. Leaves can be woven to make mats, sandals, baskets, etc. The stems provide a good coil foundation for baskets. Additionally, the stalks have been used as arrows and hand drills. A toothbrush can be fashioned from the fuzzy stem with the flowers removed.

Medicinally, the chopped or pounded rhizome was applied to the skin for minor wounds and burns (Willard 1992). Cattail down was used as dressings for wounds. Brown (1983) indicates that a sticky juice derived from between the young leaves can be used as a styptic, antiseptic, and anaesthetic. The jelly from between the young leaves can be applied to wounds, sores, external inflammations, and boils to soothe pain. Brown (1983) also indicates that the jelly was rubbed on the gums as a novocaine substitute for dental extraction.

GLOSSARY

To Medical Terms

Acrid. Sharp, irritating or biting to the taste.

Alkaloid. A nitrogen containing, slightly alkaline substance that is often poisonous.

Alterative. A substance that gradually restores the normal functions of the body.

Analgesic. Relieves pain.

Anaphrodisiac. An agent that reduces sexual desire or potency

Anesthetic. A substance that produces anesthesia

Anodyne. Helps to quiet or relieve pain.

Antibiotic. Helps to destroy pathogenic action of microbes.

Anti-inflammatory. To reduce or neutralize inflammation.

Anti-microbial. Something that inhibits the growth or multiplication of microorganisms, or kills them

Anti-pyretic. Against fever

Antiscorbutic. To be used against scurvy.

Antiseptic. Prevents infection.

Antispasmodic. Relieves or cures spasms or irregular and painful action of the muscles (e.g., epilepsy).

Anti-syphilitic. Relieves or cures venereal disease

Anti-viral. An agent used against viruses

Astringent. To shrink or bind tissue.

Bitters. Sharp acrid or biting medicines, prescribed to stimulate an appetite.

Cardiacs. To have an effect on the heart.

Carminatives. To dispel flatulency or griping pains of the stomach and bowels.

Cathartics. To stimulate the action of the bowels, a purgative.

Decoction. The essence of a plant extracted by boiling it down.

Demulcent. To have a soothing or emollient effect on inflamed surfaces.

Dermititis. Inflammation of the skin

Diaphoretics. To promote or increase perspiration.

Diuretics. To increase the flow of urine, by acting on the kidneys.

Emetics. Something that causes vomiting.

Emollients. To have a soothing and softening effect on the body tissues.

Expectorants. To cause an increase in expectoration, promoting the excretion of mucous from the chest.

Febrifuges. Help reduce or control fevers.

Glycocides. Any of the numerous acetal derivatives of sugars that on hydrolysis yield a sugar

Hydrocarbons. An organic compound containing one hydrogen and one carbon and often occurring in petroleum, natural gas, and coal.

Infusion. A preparation made by soaking a plant in hot water ("tea").

Insecticidal. Used against insects

Laxatives. Used to loosen the bowels and relieves constipation.

Liniment. A liquid or semi-liquid preparation of a herb to relieve skin irritation and muscle pain.

Mucilaginous. Something that resembles or contains mucilage (slimy).

Narcotics. Diminish the action of the nervous and vascular systems, causing drowsiness, lethargy, stupor, and insensibility.

Nervines. Soothe and calm the nerves, restoring them to a natural state

Nitrates. A salt or ester of nitric acid

Nutritives. Nourish the body, promoting growth or health.

Panacea. A "cure-all"

Parch. To toast or scorch with heat.

Photosensitivity. Being sensitive to light

Poultice. A moist, usually warm or hot mass of plant material applied to the skin, or with a cloth between the skin and plant material, to effect a medicinal action.

Purgatives. To evacuate the bowels, but more forcefully than a laxative.

Salve. A healing ointment

Saponin. A glycoside in plants that when shaken with water has a foaming or soapy action.

Sedative. Something used to lessen nervous excitement, irritation, and pain.

Selenium. A non-metalic element that resembles sulfur chemically.

Stimulants. Something produces energy.

Styptics. Helps control bleeding by contracting the tissues or blood vessels.

Sudorfic. To produceprofuse and visible sweating when taken hot

Tincture. A diluted alcohol solution of plant parts.

Tonic. To invigorate or stimulate, producing a feeling of well-being or strength.

Topical. Local application.

Vasoconstrictor. An agent that causes the blood vessels to constrict.

Vasodialator. An agent that causes the blood vessels to dialate.

Vermifuges. To expel worms or other parasites from the body.

Volatile. Readily vaporizes at low temperatures

Vulneraries. Used in the healing of wounds.

To Botanical Terms

Achene. Small, dry and hard, 1 celled, 1-seeded fruit that is indehiscent

Acute. Sharp pointed

Alpine. Occurring above treeline in the mountains.

Alternate. Arranged with one structure (e.g., leaf, flowers, stem, etc.) per node.

Angiosperm. A plant producing flowers and bearing ovules (seeds) in an ovary (fruit).

Annual. A plant, usually with a slender taproot, completing its life cycle in a single growing season.

Anther. The pollen bearing portion of the stamen.

Aquatic. Growing in water.

Areola. Small defined area on a surface of a cactus that bears the spines

Aromatic. Having a strong, usually agreeable odor.

Asexual. Without sex

Awn. A slender bristle-like organ

Axil. The upper angle formed by a leaf or branch with the stem

Banner. The upper petal of a papilionaceous flower.

Basal. At the base

Beak. A prolonged, slender, and tapering projection.

Berry. A fleshy or pulpy fruit developed from a single ovary with more than one seed, such as a grape or blueberry.

Biennial. A plant completing its life cycle in two growing seasons; usually forming a basal rosette the first season and flowering the second.

Borne. Produced or arising from

Bract. A leaf subtending a flower or flower cluster.

Bulb. An underground organ constituted mostly of fleshy storage leaves and scale covered (e.g., onion).

Calyx. A flower's sepals considered as a unit.

Campanulate. Bell-shaped

Capillary. Hair- or thread-like.

Capsule. A dry dehiscent fruit, composed of more than one carpel

Carpel. A modified leaf forming the ovary.

Catkin. In plants such as willows, birches and alders, the elongated, pendulous or conelike flower cluster with minute flowers that lack, or almost, the petals and sepals.

Caudex. Thickened base of some perennial herbs

Clasping. Partly surrounding the stem.

Clavate. Club-shaped

Claw. The narrow or stalk-like base of some petals

Coma. The tufts of hairs at the ends of some seeds

Compound Leaf. One which is divided into two or more distinct leaflets.

Cone. A dense cluster of modified, leaflike organs bearing pollen, spores, or seeds as in horsetails, clubmosses, and conifers (e.g., pinecone).

Conifer. A cone bearing tree.

Coniferous. Having cones or strobili.

Corm. A bulb-like underground thickening of the stem.

Corolla. The petals considered as a unit, usually brightly colored.

Corymb. Convex or flat-topped flower cluster of the racemose type, with the pedicels arising from a different point on the axis.

Cruciform. Cross-shaped (e.g., the position of petals in the mustard family: Brassicaceae).

Cyathium. The inflorescence in the genus *Euphorbia* (Family Euphorbiaceae), that consists of unisexual flowers crowded within a cup-like involucre.

Cyme. A flower cluster, often flat-topped or convex, in which the central or terminal flower blooms the earliest.

Deciduous. Falling off once a year, usually at the end of a growing season.

Dehiscent. Opening to emit the contents

Dichotomous. Forking regularly by pairs

Discoid. A flowering head without ray flowers

Disk. In the sunflower family (Asteraceae), the central portion of the head that gives rise to the disk flowers.

Disk flowers. In the sunflower family (Asteraceae), the flowers with slender, tubular corollas at the central part of the head.

Divided. Cut or lobed to the base or to a midrib.

Drupe. A fleshy 1-seeded fruit (e.g., cherry).

Emergent. With the lower portion in water, and the upper portion extending out.

Endemic. Found only within a limited geographic area.

Entire. With margins not cut, cleft, or otherwise toothed.

Ephemeral. Lasting for only a short time.

Evergreen. Retaining leaves through the winter.

Exotic. Not native, but introduced from somewhere else.

Fascicle. Small cluster of leaves, flowers, etc.

Family. Group of related genera.

Flower. The reproductive portion of the plant consisting of stamens, pistils, or both, and including the petals, sepals, or both.

Foliage. The leaves of the plant, collectively

Follicle. A fruit consisting of a single carpel, dehiscing by the ventral suture

Forb. A non-grasslike herbaceous plant

Frond. The leaf of a fern.

Fruit. The mature ovary, that includes the attached external structures and enclosed seeds.

Genus. A grouping of related species. The plural, of which is genera.

Glabrous. Devoid of hairs

Gland. A secreting cell or group of cells

Glandular. Having glands, usually hairs.

Glaucous. Covered or whitened with a bloom

Glochid. A barbed hair or bristle (e.g., as in the fine hairs of *Opuntia*).

Gymnosperm. A member of the plant group that characterized as having ovules not enclosed in an ovary (e.g., pines, spruces, firs, junipers).

Habit. The general appearance or growth form of a plant.

Habitat. The environmental conditions or kind of place in which a plant grows.

Head. A type of inflorescence with mostly sessile flowers densely set on a very short axis or disk, thereby having a round outline. The terminal collection of flowers surrounded by an involucre, as in the sunflower family (Asteraceae).

Herb. A plant with the aerial portion being non-woody, dying back to the ground at the end of the growing season.

Herbaceous. Not woody, dying back at the end of the growing season.

Host. The plants from which the parasite obtains nutrients (e.g., mistletoe)

Immersed. Growing under water.

Indehiscent. Not splitting open

Inferior. Lower or below

Inferior ovary. Ovary positioned below the base of other flower parts

Inflated. Turgid and bladdery

Inflorescence. The flowering part of plants; the arrangement of flowers

Inner bark. The cambium layer

Irregular. A flower where one or more of the organs are unlike the rest

Involucre. A whorl of bracts subtending a flower or flower cluster.

Latex The milky sap of certain plants.

Leaflet. One of the divisions of a compound leaf.

Legume. A simple dry fruit that is dehiscent along both sutures

Ligulate flower. The same as a ray flower in the Sunflower Family

Linear. Long and narrow

Many. For botanical purposes, numbering more than ten.

Meal. Pertaining to flour

-merous. Parted, having sections, as a 5-merous flower has 5 petals and 5 sepals.

Montane. Of or pertaining to the mountains.

Naturalized. Plants introduced from somewhere else, and now established.

Node. A joint or point of origin for leaves or branches.

Numerous. In botanical terms, more than ten.

Nut. A dry, hard walled and indehiscent fruit, usually with one seed.

Nutlet. A small nut.

Opposite. Nodes having two leaves or branches directly across from each other.

Orbicular. Circular in outline

Pappus. In the sunflower family, the highly modified calyx composed of scales, bristles, awns, or short crown at the tip of the achene.

Parasitic. Growing on and deriving nourishment from another living plant.

Parted. Cleft almost to the base

Pedicel. Stalk of a single flower

Perennial. Plant with the potential to live more than 2 years.

Perianth. Corolla and calyx considered collectively.

Petal. A member of the whorl of floral organs, just interior the sepals and below the stamens.

Petaloid. Brightly colored and petal-like.

Petiole. Leaf stalk

Pinnate. Having a main central axis with secondary branches or units arranged in two lines on either side of the central axis.

Pinole. The flour from various seeds and grains mixed together.

Pistil. Organ formed from the combination of the stigma, style, and ovary.

Plumose. Feathery and soft.

Pod. Any kind of dry, dehiscent fruit, particularly in the pea family (Fabaceae).

Pollen. Dustlike cells produced in the anther.

Pollinium. A mass of waxy or coherent pollen grains (e.g., notably in the Asclepiadaceae and Orchidaceae families).

Potherb. Herb that is boiled and eaten as a vegetable

Prostrate. Lying flat on the ground

Pubescent. Covered with hairs

Raceme. Inflorescence with one main axis and subequal primary branches each bearing one flower.

Ray flower. In the sunflower family (Asteraceae), the straplike flowers attached to the disk.

Receptacle. End of a flower stalk that bears the floral organs

Regular. Having members of each part alike in size and shape

Rhizomatous. Possessing rhizomes.

Rhizome. A creeping underground, usually horizontally oriented stem.

Root. The underground part of a plant.

Rootstalk. Underground, creeping stem.

Rosette. A dense, usually basal, cluster of leaves radiating in all directions from the stem.

Sagittate. Arrowhead shaped.

Salverform. A corolla with a long slender tube, abruptly flaring into a circular limb.

Samara. A winged fruit that does not split at maturity.

Saprophyte. A plant with little or no chlorophyll that obtains nutrients from dead organic matter by a root association with a fungus.

Schizocarp. Dry indehiscent fruit that splits into separate one-seeded segments at maturity.

Seed. A mature ovule which following germination gives rise to a new plant.

Seed cone. The female seed producing cone of conifers.

Sepal. One of a whorl of typically green or greenish, leaflike, floral organs originating below the petals.

Septum. A partition that seperates the locules of an ovary.

Serrate. Sharply toothed edges

Sessile. Lacking a stalk.

Shoot. Young stem or branch

Shrub. A woody plant, sometimes only at the base, and generally with several stems originating from the base.

Silicle. Dry, dehiscent fruit of the Mustard Family, typically less tha twice as long as wide

Silique. Dry, dehiscent fruit of the Mustard Family, typically more than twice as long as wide.

Spatulate. Spatula shaped; having a long narrow base and a widened, roundish tip.

Spore. A single cell or a small group of undifferentiated cells, each capable of producing a plant

Spur. A hollow slender sac-like extension of some part of the flower (e.g., sepal in *Delphinium* or the petal of *Viola*).

Stamen. The male organ of a flower that produces pollen. It is composed of an anther and filament.

Stellate. Star shaped

Stem. The main axis of a plant.

Stipules. Appendages on each side of the base of certain leaves

Subalpine. Growing in the mountains below the alpine zone and above the montane zone.

Succulent. Thick, fleshy, and juicy.

Superior. Above

Superior ovary. Ovary positioned above the base of the other flower parts

Talus. Slope of rock rubble, usually at a cliff base.

Taproot. The primary plant root that considerably larger than any other root system branches.

Tepal. The perianth part when the perianth is not clearly differentiated into calyx or corolla.

Terrestrial. Growing on ground, not aquatic.

Three-ranked. Originating in threes from a common point or level.

Tree. Large woody plant with a single main stem or trunk.

Tuber. A swollen underground stem tip (e.g., potato).

Tubular. In the form of a tube or cylinder.

Umbel. Flower arrangement resembling an "umbrella"

Urn shaped. Ovoid and with a small opening at the tip.

Vascular plant. A plant having vascular tissue.

Vegetative. The portion of the plant not producing reproductive structures like cones or flowers.

Villous. Sticky

Weed. An aggressive plant that colonizes disturbed habitats and cultivated lands. A plant out of place.

Whorled. Three or more similar structures (e.g., leaves, petals, bracts) encircling a node

Woolly. Having soft, curled or entangled hairs.

REFERENCES AND ADDITIONAL READING

Alexander, R.R. 1985. Major habitat types, community types, and plant communities in the Rocky Mountains. U.S. Forest Service General Technical Report RM-123.

Allen, E.B. and D.H. Knight. 1984. The effects of introduced annuals on secondary succession in sagebrush-grassland. Wyoming Southwest. Natural 29:407-421

Altschul, S. 1973. Drugs and Food from little-known Plants. Harvard University Press, Cambridge, Massachusetts.

Anderson, J.P. 1939. Plants used by the Eskimo of the Northern Bering Sea and Arctic Regions of Alaska. American Journal of Botany 26(9):714-716.

Angier, B. 1966. Free for the Eating. Stackpole Books, Harrisburg, Pennsylvania

Angier, B. 1969a. More Free for the Eating Wild Foods. Stackpole Books, Harrisburg, Pennsylvania.

Angier, B. 1969b. Feasting Free on Wild Edibles. Stackpole Books, Harrisburg, Pennsylvania.

Angier, B. 1974. Field Guide to Edible Wild Plants. Stackpole Books, Harrisburg, Pennsylvania.

Anon, 1984. Traditional medicine project. Avataq Cultural Institute.

Arnason, T., R.J. Hebda and T. Johns. 1981. Use of Plants for Food and Medicine by Native Peoples of Eastern Canada. Canadian Journal of Botany 59(11): 2189 2325.

Ashton, R.J. and R.D. Walmsley. 1976. The aquatic fern *Azolla* and its *Anabaena* symbiont. Endeavour 35:39-43.

Avery, A.G., S. Satina, and J. Rietsema 1959. Blakeslee: The genus Datura. Ronald Press Company, New York

Bacon, A.E. 1903. An Experiment with the Fruit of Red Baneberry. Rhodora 5:77-79

Bailey, F.L. 1940. Navajo Foods and Cooking Methods. American Anthropologist 42(2):270-290, April-June.

Baker, M.A. 1981. The Ethnobotany of the Yurok, Tolowa and Karok Indians of Northwest California. M.A. Thesis, Humboldt State University, Arcata, California.

Balls, E.K. 1970. Early Uses of California Plants. Berkeley and Los Angeles, University of California Press.

Bank, T.P, II. 1951. Botanical and Ethnobotanical Studies in the Aleutian Islands I. Aleutian Vegetation and Aleut Culture. Botanical and Ethnobotanical Studies Papers, Michigan Academy of Science, Arts and Letters, 37: 13 30.

Barrett, S. A. 1908. Pomo Indian Basketry. University of California Publica¬tions in American Archaeology and Ethnology 7:134 308.

Barrett, S. A. 1917. The Washoe Indians. Bulletin of the Public Museum of the City of Milwaukee 2(1): 1 52.

Barrett, S. A. 1952. Material Aspects of Pomo Culture. Bulletin of the Public Museum of the City of Milwaukee, Number 20.

Barrett, S.A. and E.W. Gifford. 1933. Miwok Material Culture. Yosemite Natural History Association, Inc., Yosemite National Park, California

Barrows, D.P. 1967. The Ethno Botany of the Coahuilla Indians of Southern California. Banning, California: Malki Museum Press. Origi¬nally published in 1900.

Bartlett, K. 1943. Edible Wild Plants of Northern Arizona. Plateau 16(1):11-17. Northern Arizona Society of Science and Art Museum of Northern Arizona, Flagstaff.

Baumhoff, M.A. 1963. Ecological Determinants of Aboriginal California Populations. University of California Publications in American Archaeology and Ethnology 49(2):155-236.

Bean, L.J. and K.S. Saubel. 1972. Temalpakh-Cahuilla Indian Knowledge and Usage of Plants. Malki Museum Press, Morongo Indian Reservation, California.

Bomhard, M.L. 1936. Leaf venation as a means of distinguishing *Cicuta* from *Angelica*. Journal of the Washington Academy of Sciences 26(3):102-107.

Boxer, A. 1974. Nature's Harvest. Henry Regnery Company, Chicago, Illinois

Brill, S. 1994. Identifying and Harvesting Edible and Medicinal Plants in Wild (and Not so Wild) Places. Hearst Books, New York.

Brown, T. 1985. Tom Brown=s Guide to Wild Edible and Medicinal Plants. Berkeley Books, New York.

Bryan, N.G. and S. Young. 1940. Navajo Dyes, Their Preparation and Use. Education Division Publication, U.S. Office of Indian Affairs, February.

Burgess, R.L. 1966. Utilization of Desert Plants by Native People. Contribution committee on desert and arid zone research. Bulletin of the American Association for the Advancement of Science 8:6-21.

Burt, P. 2000. Barrenland Beauties: showy plants of the Canadian Arctic. Outcrop Limited. The Northern Publishers, Yellowknife, N.W.T. 238 p.

Callegari, J. And K. Durand. 1977. Wild Edible and Medicinal Plants of California. El Cerrito, California

Camazine, S. and R.A. Bye 1980. A Study of the Medical Ethno-botany of the Zuni Indians of New Mexico. Journal of Ethnopharmacology 2: 365 388.

Castetter, E.F. 1935. Uncultivated Native Plants Used as Sources of Food. Ethnobiological Studies in the American Southwest. The University of New Mexico Bulletin, Whole Number 266, Biological Series Vol. 4, No. 1, University of New Mexico Press, Albuquerque.

Chamberlain, L.S. 1901. Plants used by the Indians of Eastern North America. American Naturalist, 35:1-10.

Chamberlin, Ralph V. 1911. The Ethno-Botany of the Gosiute Indians of Utah. Memoirs of the American Anthropological Association 2(5):331-405.

Chatfield, K. 1997. Medicine from the Mountains: Medicinal Plants of the Sierra Nevada. South Lake Tahoe, California: Range of Light Press.

Chestnut, V.K. 1902. Plants used by the Indians of Mendocino County, California. Contributions U.S. National Herbarium 7:295-408.

Clarke, C.B. 1977. Edible and Useful Plants of California. University of California Press, Berkeley, California

Classen, P.W. 1919. A Possible New Source of Food Supply (Cat-tail Flour). Scientific Monthly 9:179-185

Coffey, T. 1993. The History and Folklore of North American Wildflowers. Facts on File, Inc., New York

Coon, N. 1974. The dictionary of useful plants. Rodale Press/Book Division, Emmaus, Pennsylvania

Coulter, J. M. 1885. Manual of the Botany (Phaenogamia and Pteridophyta) of the Rocky Mountain Region, from New Mexico to the British Boundary. American Book Co., New York, xvi + 452 pp.

Coulter, J. M. and A. Nelson. 1909. New Manual of Botany of the Central Rocky Mountains (Vascular Plants). American Book Co., New York, 646 pp.

Coville, F.V. 1897. Notes on the plants used by the Klammath Indians of Oregon. Contributions to the U.S. National Herbarium, 5(2).

Coville, F.W. 1904. Desert Plants as a Source of Drinking Water. Smithsonian Institute Annual Report, pp. 499-505.

Craighead, J.J., F.C. Craighead, and R.J. Davis. 1963. A Field Guide to the Rocky Mountain Wildflowers. Boston: Houghton Mifflin Company.

Cronquist, A., A. H. Holmgren, N. H. Holmgren, and J. L. Reveal. 1972. Intermountain Flora: Vascular plants of the Intermountain West, U.S.A. Hafner Publishing Co., Inc., New York, Vol. 1, iii + 270 pp.

Culley, D.D. and E.A. Epps. 1973. Use of duckweed for waste treatment and animal feed. Journal of Water Pollution Control Federation 45:337-347

Culpeper, N. 1972. English Physician and Complete Herbal. Arranged for use as a First Aid Herbal, by leyel, C.F. No. Hollywood, California. Wilshire Book Company.

Curtin, L. S. M. 1957. Some Plants Used by the Yuki Indians ... I. Historical Review and Medicinal Plants. Masterkey 31: 40 48.

Dall, W.H. 1868. Useful Indigenous Alaskan Plants. Report of the Department of Agriculture, USDA, Washington D.C. pp. 172-189.

Darlington, W. 1859. American Weeds and Useful Plants. New York, A.O. Moore

Davidson, J. 1919. Douglas-fir Sugar. The Canadian Field Naturalist 33(1)6-9.

Dawson, R. 1985. Nature Bound. OMNIgraphics Ltd., Boise, Idaho

Densmore, F. 1974. How Indians Use Wild Plants for Food, Medicine, and Crafts. Dover Publications, New York.

Despain, D.G. 1975. Field key to the flora of Yellowstone National Park. Yellowstone Library and Museum Association, Yellowstone National Park, WY. 155 pp.

Despain, D.G. 1999. Yellowstone Vegetation. Robert Rinehart, Inc. in cooperation with The Yellowstone Association.

Doebley, J.F. 1984. Seeds of Wild Grasses: A Major Food of Southwestern Indians. Economic Botany 38(I):52-64.

Dorn, R.D. 1977. Manual of the Vascular Plants of Wyoming. Garland Publishing Inc., New York, 2 vols., 1498 pp.

Dorn, R.D. 1988. Vascular Plants of Wyoming. Mountain West Publ., Cheyenne, WY, vi + 340 pp.

Dorn, R.D. 1992. Vascular plants of Wyoming. 2nd edition. Mountain West Publishing, Cheyenne, WY. 340 pp.

Douglas, J.S. 1978. Alternative Foods: A World Guide to Lesser-known Edible Plants. Pelham Books, London

Duke, J.A. 1992. Handbook of Medicinal Plants. CRC Press, Boca Raton, Florida

Duke, J.A. 1992. Handbook of Phytochemical Constituents of GRAS Herbs and Other Economic Plants. CRC Press, Boca Raton, Florida

Dunmire, W.W. and G.D. Tierney. 1997. Wild Plants and Native Peoples of the Four Corners. Museum of New Mexico Press, Santa Fe.

Ebeling, W. 1986. Handbook of Indian Foods and Fibers of Arid America. University of California Press, Berkeley, Los Angeles, London

Elias, T.S. and P.A. Dykeman. 1982. A Field Guide to North American Edible Wild Plants. Outdoor Life Books, New York.

Elliott, D.B. 1976. Roots: An Underground Botany and Foragers Guide. The Chatham Press, Old Greenwich, Connecticut.

Elliot, D.B. Wild Roots: A Forager=s Guide to the Edible and Medicinal Roots, Tubers, Corms and Rhizomes of North America. Healing Arts Press, Rochester, Vermont.

Ellis, C. 1941. Wild Vegetables of the Desert Indians. Primitive Man, number 3: 9-10

Elmore, F.H. 1944. Ethnobotany of the Navajo. Sante Fe, New Mexi¬co: School of American Research.

Erichsen-Brown, C. 1979. Medicinal and Other Uses of North American Plants. Dover Publications, Inc., New York.

Evert, E.F. 1984a. A new species of Antennaria (Asteraceae) from Montana and Wyoming. Madrono 31: 109-112.

Evert, E.F. 1984b. Penstemon absarokensis, a new species of Scrophulariaceae from Wyoming. Madrono 31: 140-143.

Evert, E.F. and L. Constance. 1982. Shoshonea pulvinata, a new genus and species of Umbelliferae from Wyoming. Syst. Bot. 7: 471-475.

Evert, E.F. and R.L. Hartman. 1984. Additions to the vascular flora of Wyoming. Great Basin Naturalist 44: 482, 483. (ten Wyoming state records).

Farris, G. 1980. A re-assessment of the nutritional value of *Pinus monophylla*. Journal of California and Great Basin Anthropology 2(1): 132-36.

Fernald M. and A.C. Kinsey. 1958. Edible Wild Plants of Eastern North America, revised by R.C. Rollins. Harper and Row, New York.

Fertig, W. 1992a. Checklist of the Vascular Plant Flora of the West Slope of the Wind River Range and a Status Report on the Sensitive Plant Species of Bridger-Teton National Forest. Report to Bridger-Teton National Forest. viii + 255 pp.

Fertig, W. 1992b. Checklist of the Vascular Plant Flora of the West Slope of the Wind River Range and Status Report on Sensitive Plant Species Occurring in the Rock Springs District, Bureau of Land Management. Report to Rock Springs District, BLM. iv + 61 pp.

Fertig, W., R.L. Hartman, and B.E. Nelson. 1991. General Floristic Survey of the West Slope of the Wind River Range, Bridger-Teton National Forest, 1990. Report to Bridger-Teton National Forest. iii + 144 pp.

Fertig, W. 1999. A potpourri of weeds. Castilleja 18(2):3B10.

Flora of North America (FNA) Editorial Committee. 1992. Flora of North America: Guide for Contributors. Missouri Botanical Garden.

Flora of North America (FNA) Editorial Committee. 1993. Flora of North America Volume 2: Pteridophytes and Gymnosperms. Oxford University Press. New York.

Foster S. and J.A. Duke. 1990. Eastern/Central Medicinal Plants. Peterson Field Guides, Houghton Mifflin Company, New York

Fowler, C.S. 1989. Willard Z. Park's Ethnographic Notes on the Northern Paiute of Western Nevada 1933 1940. Salt Lake City: University of Utah Press.

Frankton, C. and G.A. Mulligan 1987. Weeds of Canada. N.C. Press Limited and Agricultural Canada, Ottawa, Ontario.

Frye, T.C. 1934. Ferns of the Northwest. Metropolitan Press, Portland, Oregon

Gail, F.W. 1916. Some Poisonous Plants of Idaho. University of Idaho Agricultural Experiment Station Bulletin 86.

Gibbons, E. 1962. Stalking the Wild Asparagus. New York: David McKay Company.

Gibbons, E. 1966. Stalking the Healthful Herbs. David McKay, New York.

Gibbons, E. 1971. Stalking the Good Life. David McKay, New York

Gifford, E. W 1967. Ethnographic Notes on the Southwestern Pomo. Anthropological Records 25: 10 15.

Gilmore, M.R. 1919 Uses of Plants by the Indians of the Missouri River Region. SI-BAE Annual Report #33

Goodrich, J. and C. Lawson. 1980. Kashaya Pomo Plants. Los Angeles: American Indian Studies Center, University of California, Los Angeles.

Gottesfeld, L.M. J. 1992. The Importance of Bark Products in the Abo¬riginal Economies of Northwestern British Columbia, Canada. Economic Botany 46(2): 148 157.

Grant, A.L. 1924. A Monograph of the Genus *Mimulus*. Washington University Doctoral Dissertation, Publications of Washington University Series V, St. Louis

Grillos, S.J. 1966. Ferns and Fern Allies of California. University of California Press, Berkeley, California

Gunther, E. 1973. Ethnobotany of Western Washington, revised edition. University of Washington Press, Seattle and London.

Haines, J. 1988. A Flora of the Wind River Basin and adjacent areas, Fremont, Natrona, and Carbon Counties. M.S. Thesis, University of Wyoming, ii + 76 pp.

Hall, A. 1976. The Wild Food Trail Guide. Holt, Rinehart, and Winston, New York

Hamel, P.B. and M.U. Chiltoskey. 1975. Cherokee Plants and Their Uses A 400 Year History. Sylva, North Carolina: Herald Publishing.

Hardin, J.W. and J.M. Arena. 1974. Human Poisoning from Native and Cultivated Plants. Duke University Press, Durham, North Carolina.

Harrington, H.D. 1967. Edible Native Plants of the Rocky Mountains. Albuquerque: University of New Mexico Press.

Hart, J.A. 1996. Montana - Native Plants and Early Peoples. Montana Historical Society, Helena, Montana.

Hart, J.A. 1981. The Ethnobotany of the Northern Cheyenne Indians of Montana. Journal of Ethnopharmacology 4:1-55.

Hartman, R.L. 1990. The flora of the Rocky Mountains project. Amer. Jour. Bot. 77:(6): 134, 135.

Hartman, R. L. and R. W. Lichvar. 1980. Additions to the vascular flora of Teton County, Wyoming. Great Basin Naturalist 40: 408-413. (125 taxa new to Teton County)

Harvard, V. 1895. The Food Plants of North American Indians. Bulletin of Torrey Botanical Club 22:98-123.

Haskin, L.L. 1929. A Frontier Food, Ipo or Yampa, Sustained the Pioneers. Nature Magazine 14:171-172

Hedges, K. 1986. Santa Ysabel Ethnobotany. San Diego Museum of Man Ethnic Technology Notes, Number 20.

Hellar, C.A. 1958. Wild Edible and Poisonous Plants of Alaska. University of Alaska Extension Bulletin, no. 40.

Heller, C.A. 1966. Wild, edible and poisonous plants of Alaska. Division of Statewide Services, Cooperative Extension Service, University of Alaska.

Hellson, J.C. 1974. Ethnobotany of the Blackfoot Indians. National Museums of Canada, Ottawa, Ontario.

Herrick, J.W. 1977. Iroquois Medical Botany. Ph.D. Thesis, State University of NewYork, Albany.

Hill, A.F. 1937. Economic Botany: A Textbook of Useful Plants and Plant Products. McGraw-Hill Book Company, New York.

Hinton, L. 1975. Notes on La Huerta Diegueno Ethnobotany. Journal of California Anthropology 2: 214 222.

Hitchcock, C.L, A. Cronquist, M. Ownbey, and J.W. Thompson. 1955. Vascular Plants of the Pacific Northwest part 5. Univ. Wash. Publ. Biol. 17.

Hitchcock, C.L, A. Cronquist, M. Ownbey, and J.W. Thompson. 1959. Vascular Plants of the Pacific Northwest part 4. Univ. Wash. Publ. Biol. 17.

Hitchcock, C.L, A. Cronquist, M. Ownbey, and J.W. Thompson. 1961. Vascular Plants of the Pacific Northwest part 3. Univ. Wash. Publ. Biol. 17.

Hitchcock, C.L, A. Cronquist, M. Ownbey, and J.W. Thompson. 1964. Vascular Plants of the Pacific Northwest part 2. Univ. Wash. Publ. Biol. 17.

Hitchcock, C.L, A. Cronquist, M. Ownbey, and J.W. Thompson. 1969. Vascular Plants of the Pacific Northwest part 1. Univ. Wash. Publ. Biol. 17.

Hocking, G.M. 1949. From Pokeroot to Penicillin. Rocky Mountain Druggist, November 1949, pages 12, 38.

Holloway, P. and G. Alexander. 1990. Ethnobotany of the Fort Yukon Region, Alaska. Economic Botany 44:214-225.

Holt, Catharine. 1946. Shasta Ethnography. Anthropological Records 3(4): 308.

Hough, W. 1897. The Hopi in Relation to Their Plant Environment. The American Anthropologist 10(2):33-44.

Haughton, C. S. 1978. Green Immigrants. Harcourt Brace Jovanovich, New York

Hussey, P.B. 1939. A Taxonomic List of Some Plants of Economic Importance. The Science Press Printing Company, Lancaster, Pennsylvania.

Jacobson, C.A. 1915. AWater Hemlock (Cicuta).@ Nevada Agricultural Experiment Station, Technical Bulleton #81.

Jencks, Z. 1919. A Note on the Carbohydrates of the Root of the Cattail (Typha latifolia). Proceedings of the Society for Experimental Biology and Medicine 17(2):45-46, November 19.

Johnston, A. 1970. Blackfoot Indian Utilization of the Flora of the Northwestern Great Plains. Economic Botany 24:301-324.

Kartesz, J. 1994. A Synonymized Checklist of the Vascular Flora of the United States, Canada, and Greenland, volumes 1 and 2. Timber Press. Portland, OR.

Kavash, B. 1979. Native Harvests: recipes and botanicals of the American Indian. First Vintage Books, A Division of Random House, New York.

Kelly, I. 1932. Ethnography of the Surprise Valley Paiute. University of California Publications in American Archaeology and Ethnology 31(3): 67 210.

Kephart, H. 1909 and later editions. The Book of Camping and Woodcraft, Chapter 17, AEdible Plants of the Wilderness.@ The Century Company, New York

Kindscher, K. 1987. Edible Wild Plants of the Prairie. University Press of Kansas.

Kinghorn, D. 1979. Toxic Plants. Columbia University Press, New York.

Kingsbury, J.M. 1964. Poisonous Plants of the United States and Canada. Englewood Cliffs, New Jersey: Prentice-Hall.

Kingsbury, J.M. 1965. Deadly Harvest: A Guide to Common Poisonous Plants. New York: Holt, Rinehart, and Winston.

Kirk, D.R. 1975. Wild Edible Plants of the Western United States. Healdsburg, California: Naturegraph Publishers.

Kirkpatrick, R. S. 1987. A Flora of the Southeastern Absarokas, Wyoming. M.S. Thesis, University of Wyoming, vi + 166 pp.

Knap, A.H. 1975. Wild Harvest: An Outdoorsman=s Guide to Edible Wild Plants in North America. Pagurian Press, Ltd., Toronto, Canada

Knight, D.H. 1994. Mountains and Plains: Ecology of Wyoming Landscapes. Yale University Press, New Haven and London

Krochmal, A. and C. Krochmal. 1973. A Guide to the Medicinal Plants of the United States. Quadrangle/New York Times Book Company.

Krochmal, A, S. Paur, and P. Duisberg. 1951. Useful Native Plants in the American Deserts. Economic Botany 8(1):3-20.

Lackschewitz, K. 1991. Vascular Plants of West-central Montana - Identification Guidebook. General Technical Report INT-277. Ogden Utah: U.S. Department of Agriculture, Forest Service, Intermountain Research Station. 648p.

Lands, M. 1959. Folk Medicine and Hygiene. Anthropological Papers of the University of Alaska 8: 1 75.

Lee, D. 1989. Exploring Nature=s Uncultivated Garden. Havelin Communications, Tacoma park, Maryland.

Le Strange, R. 1977. A History of Herbal Plants. Arco Publishing Company, Inc., New York

Lewis, W.H. and M. Elvin Lewis. 1977. Medical Botany. John Wiley and Sons, New York.

Life-Support Technology, Inc. 1963. Foods in the Wilderness.

Linn, J.G , E.J. Sraba, R.D. Goodrich, J.C. Meiske, and D.E Otterby. 1975. Nutritive value of dried and ensiled aquatic plants. I. Chemical composition. Journal of Animal Science 41:601-609

Lichvar, R. W. 1979a. The Flora of the Gros Ventre Mountains. M.S. Thesis, University of Wyoming, 384 pp.

Lust, J.B. 1987. The Herb Book, 20th Edition. Bantam Books, New York

Mabey, R. 1977. Plantcraft: A Guide to Everyday Use of Wild Plants. Universe Books, New York

Mahar, J.M. 1953. Ethnobotany of the Oregon Paiutes of the Warm Springs Indian Reservation. B.A. Thesis, Reed College, Portland, Oregon.

Markow, S. 1992. Preliminary report on a general floristic survey of vascular plants of the Targhee National Forest. Report to Teton National Forest, 39 pp.

Marriott, H. 1985. Flora of the Northwestern Black Hills, Crook and Weston Counties, Wyoming. M.S. Thesis, University of Wyoming, iii + 93 pp.

Martin, L.C. 1984. Wildflower Folklore. Chester, Connecticut: The Globe Pequot Press.

McHarg, I.L. 1969. Design with Nature. Garden City, New York. Doubleday

Medsger, O.P. 1974. Edible Wild Plants. Collier-Macmillan Publishers, New York

Meriam, C.H. 1918. The Acorn, a Possibly Neglected Source of Food. National Geographic Magazine 34(2):129-137.

Merriam, C.H. 1966. Ethnographic Notes on California Indian Tribes. Berkeley: University of California Archaeological Research Facility.

Merrill, R.E. 1923. Plants Used in Basketry by the California Indians. UC-PAAE 20:215-242.

Meuninck, J. 1988. The Basic Essentials of Edible Wild Plants and Useful Herbs. ICS Books, Merrillville, Indiana.

Miller, J.A. 1973. Naturally occurring substances that can induce tumors, in Toxicants Occurring Naturally in Foods. National Academy of Sciences, Washington D.C.

Millspaugh, C.F. 1974. American Medicinal Plants, An Illustrated and Descriptive Guide to Plants Indigenous to and Naturalized in the United States Which are Used in Medicine. Dover Publications, New York.

Mitchell, R.S. and J.K. Dean. 1982. Ranunculaceae (Crowfoot Family) of New York State. New York State Museum Bulletin no. 446:1-100

Moerman, D.E. 1977. American Medical Ethnobotany: a reference dictionary. Garland Publishing, Inc. New York and London.

Moerman, D. 1986. Medicinal Plants of the Native Americans. University of Michigan Museum of Anthropology Technical Report, Number 19. University of Michigan, Ann Arbor.

Moore, M. 1979. Medicinal Plants of the Mountain West. Museum of New Mexico Press, Santa Fe, New Mexico

Moore, M. 1989. Medicinal Plants of the Desert and Canyon West. Museum of New Mexico Press, Santa Fe, New Mexico

Morton, J.F. 1975. Cattails (Typha spp.) - weed problem or potential crop? Econominc Botany 29(1):7-29.

Morton, J. 1963. Principal Wild Food Plants of the United States, excluding Alaska and Hawaii. Economic Botany 17:319-330.

Muenscher, W.C. 1962. Poisonous Plants of the United States. The MacMillan Company, New York

Murphey, E.V.A. 1990. Indian Uses of Native Plants. Meyerbooks, Glenwood, Illinois.

National Park Service. 1986. Exotic vegetation management plan. Yellowstone National Park, WY. 14 pp.

National Research Council. 1985. Amaranth: Modern Prospects for an Ancient Crop. Rodale Press, Emmaus, Pennsylvania

Nelson, R.A. 1992. Handbook of Rocky Mountain Plants, 4th edition. Roberts Rinehart Publishers, Niwot, Colorado.

Nequakewa, E. 1943. Some Hopi Recipes for the Preparation of Wild Plant Foods. Plateau, no. 18:18-20

Newberry, J.S. 1887. Food and Fiber Plants of the North American Indians. Popular Scientific Monthly 32:31-46.

Norton, C. 1942. Would You Starve? Nature Magazine 35(6):295-297.

Norton, H. 1981. The association between anthropogenic prairies and important food plants in western Washington. Northwest Anthropological Research Notes 13: 175-200

Olsen, L.D. 1990. Outdoor Survival Skills. Brigham University Press, Provo, Utah.

Oswalt, W H. 1957. A Western Eskimo Ethnobotany. Anthropological Papers of the University of Alaska 6: 17 36.

Palmer, E. 1878. Plants Used by the Indians of the United States. The American Naturalist 12:593-606 (Sept.) and 646-655 (Oct.).

Perry, E 1952. Ethno Botany of the Indians in the Interior of British Columbia. Museum and Art Notes 2(2): 36 43.

Peterson, L.A. 1978. A Field Guide to Edible Wild Plants of Eastern and Central North America. Peterson Field Guides, Houghton Mifflin Company, New York.

Pfeiffer, N.E. 1922. Monograph of the Isoetaceae. Annuals Missouri Botanical Gardens 9:79-232.

Pojar, J. and A. MacKinnon, (eds.). 1994. Plants of the Pacific Northwest Coast. Lone Pine Publishing, Washington.

Porsild, A.E. Edible Plants of the Arctic. Arctic 6, no. 1:15-34

Powers, S. 1874. Aboriginal Botany. Proceedings of the California Academy of Science 5: 373 379.

Price, L.W. 1981. Mountains and Man. Berkeley: University of California Press

Reagan, A.B. 1929. Plants Used by the White Mountain Apache Indians of Arizona. Wisconsin Archeologist 8:143-61.

Reagan, A. 1934. Various Uses of Plants by West Coast Indians. Wash¬ington Historical Quarterly 25: 133 137.

Risk, P. 1983. Outdoor Safety and Survival. New York: John Wiley

Ritchie, G.A., (ed). 1979. New Agricultural Crops. Westview Press, Boulder, Colorado

Rogers, D.J. 1980 Lakota Names and Traditional Uses of Native Plants by Sicangu (Brule) People in the Rosebud Area, South Dakota. St. Francis, SD. Rosebud Educational Scoiety

Romero, J.B. 1954. The Botanical Lore of the California Indians. New York: Vantage Press.

Rydberg, P.A. 1900. Catalogue of the flora of Montana and the Yellowstone National Park. Memoirs of the New York Botanical Garden. Volume 1. 491 pp.

Rydberg, P.A. 1917. Flora of the Rocky Mountains and Adjacent Plains: Colorado, Utah, Wyoming, Idaho, Montana, Saskatchewan, Alberta, and Neighboring Parts of Nebraska, South Dakota, North Dakota, and British Columbia. Published by the author, xii + 1110 pp.

Saunders, C.F. 1976. Edible and Useful Wild Plants of the United States and Canada. Dover Publications, New York.

Sawyer, J.O. and T. Keeler-Wolf. 1995. A Manual of California Vegetation. Sacramento: California Native Plant Society.

Schery, R.W. 1972. Plants for Man. 2nd ed. Prentice Hall, Englewood Cliffs.

Schofield, J.J. 1989. Discovering Wild Plants: Alaska, Western Canada, the Northwest. Alaska Northwest Books, Seattle, Washington

Scott, R. W. 1966. The Alpine Flora of Northwestern Wyoming. M.S. Thesis, University of Wyoming, v + 219 pp.

Scully, V. 1970. A Treasury of American Indian Herbs: Their Lore and Their Use for Food, Drugs, and Medicine. Crown Publishers, Inc., New York

Shaw, R. J. 1976. Field Guide to the Vascular Plants of Grand Teton National Park and Teton County, Wyoming. Utah State University Press, Logan, xvi + 301 pp.

Smith, C.E. 1973. Man and His Foods: studies in the ethnobotany and nutrition - contemporary, primitive, and prehistoric non-European diets. The University of Alabama Press, Alabama.

Smith, H.H. 1923. Ethnobotany of the Menomini Indians. Bulletin of the Public Museum of the City of Milwaukee, vol. 4, no. 1 (reprinted 1978)

Snow, C.R. 1935. Vegetables of the Alaska Wilderness. The Alaska Sportsman 1(4):6-8

Snow, N. 1989. Floristics of the Headwaters Region of the Yellowstone River (Wyoming). M.S. Thesis, University of Wyoming, viii + 116 pp.

Snow, N. 1990. Phytogeographical affinities of the Absaroka Range (Wyoming-Montana). Proc. Southwest. and Rocky Mtn. Div., AAAS meeting, Colorado Springs, CO, Program and Abstracts p. 30.

Snow, N., B.E. Nelson, and R. L. Hartman. 1990. Additions to the vascular flora of Yellowstone National Park, Wyoming. Madrono 37: 214-216.

Sparkman, P.S. 1908. The Culture of the Luiseiio Indians. University of California Publications in American Archaeology and Ethnology 8(4): 187 234.

Spellenberg, Richard 1979. The Audobon Society Field Guide to North American Wildflowers. New York, NY. A Borzoi Book published by Alfred A. Knopf.

Spier, L. 1930. Klamath Ethnography. University of California Publica¬tions in American Archaeology and Ethnology 30: 1 338.

Steward, J.H. 1933. Ethnography of the Owens Valley Paiute. University of California Publications in American Archaeology and Ethnology, 33(3): 233-350, University of California Press, Berkeley.

Stewart, Kenneth M. 1965. Mohave Indian Gathering of Wild Plants. Kiva 31(l):46 53.

Strike, S.S. 1994. Etnobotany of the California Indians, vol. 2. Aboriginal Uses of California=s Indigenous Plants. Koeltz Scientific Books, Champaign, Illinois

Stuart, J.D. and J.O. Sawyer. 2001. Trees and Shrubs of California. California Natural History Guides #62. University of California Press, Berkeley and Los Angeles, California.

Swartz, B. K., Jr. 1958. A Study of Material Aspects of Northeastern Maidu Basketry. Kroeber Anthropological Society Publications 19: 67 84.

Sweet, M. 1976. Common Edible and Useful Plants of the West. Naturegraph Publishers, Inc., Heraldsburg, California

Taylor, S. J. Elk. 1994. Eat the Weeds at Your Feet: an edible plant guide of Sonoma County. Rose of Sharon Press, Citrus Heights, California

Teit, J.A. 1930. Ethnobotany of the Thompson Indians of British Columbia. U.S. Government Printing Office, Washington, D.C.

Terrell, E.E. 1977. A Checklist of Names for 3,000 Vascular Plants of Economic Importance. U.S. Department of Agriculture Handbook No. 505, pp. 21-22.

Thompson, S. and M. Thompson. 1972. Wild Plant Foods of the Sierra. Dragtooth Press, Berkeley, California

Thompson, S. and M. Thompson. 1977. Huckleberry Country: Wild Food Plants of the Pacific Northwest. Berkeley: Wilderness Press.

Tilford, G.L. 1993. The EcoHerbalist=s Fieldbook: Wildcrafting in the Mountain West. Mountain Weed Publishing, Conner, Montana.

Tilford, G.L. 1997. Edible and Medicinal Plants of the West. Mountain Press Publishing Company, Missoula, Montana.

Train, P., J.R. Henriches, and W.A. Archer. 1957. Medicinal Uses of Plants by Indian Tribes of Nevada. Quarterman Publication, Lawrence, Massachusetts

Truax, R.E., D.D. Culley, M. Griffith, W.A Johnson, and J.P Wood. 1972. Duckweed for chick feed. Louisiana Agriculture 16(1):8-9.

Tull, D. 1987. A Practical Guide to Edible and Useful Plants. Texas Monthly Press, Austin, Texas

Turner, N. And H.V. Kuhnlein. 1991. Traditional Plant Foods of Canadian Indigenous Peoples. Gordon and Breach Science Publishers, Philadelphia, Pennsylvania.

Turner, N.J. and A.F. Szczawinski. 1991. Common Poisonous Plants and Mushrooms of North America. Timber Press, Portland, Oregon

Turney-High, H. 1933. Cooking Camass and Bitterroot. Scientific Monthly 36:262-263.

Tweedy, F. 1886. Flora of the Yellowstone National Park. Washington, DC. 78 pp.

Tyler, V.E. 1987. The Honest Herbal. George F. Stickley Company, Philadelphia, Pennsylvania

Underhill, J.E. 1974. Wild Berries of the Pacific Northwest. Seattle, Washington, Superior Publishing Company.

Uphof, J.C.T. 1959. Dictionary of Economic Plants. H.R. Engleman

Usher, G. 1976. A Dictionary of Plants Used by Man. Constable, London.

Van Etten, C.H., R.W. Miller, I.A. Wolff, and Q. Jones. 1963. Amino Acid composition of seeds from 200 angiosperm plant species. Journal of Agriculture and Food Chemistry 11(5):399-410.

Vestal, P.A. 1952. Ethnobotany of the Ramah Navaho. Papers of the Peabody Museum of American Archaeology and Ethology, Harvard University Vol. XL-No. 4

Vizgirdas, R. 1999a. The Fallacy of Plant Edibility Tests. Wilderness Way Magazine, vol. 5, issue 2.

Vizgirdas, R. 1999b. Fireweed. Wilderness Way Magazine, vol. 4, issue 4.

Vizgirdas, R. 1999c. Courting the Conifers: the Pines. Wilderness Way Magazine, vol. 4, issue 1.

Vizgirdas, R. 2000a. The Mustard Family. Wilderness Way Magazine, vol. 6, issue 1.

Vizgirdas, R. 2000b. Butterflies and Edible Plants. Wilderness Way Magazine, vol. 5, issue 3.

Vizgirdas, R.S. 2003a. Useful Plants of Idaho. Idaho State University Press, Pocatello, Idaho.

Vizgirdas, R.S. 2003b. Useful Plants of the Southern California Mountains. San Bernardino County Museum Association Quarterly 50(2). Redlands, California

Vizgirdas, R.S. and E.M. Rey-Vizgirdas. 2006a. Wild Plants of the Sierra Nevada. University of Nevada Press, Reno.

Vizgirdas, R.S. 2006b. Discovering Sawtooth's Butterflies. Idaho State University Press, Pocatello, Idaho.

Walker, M. 1984. Harvesting the Northern Wild: A guide to traditional and contemporary uses of edible forest plants of the Northwest Territories. Yellowknife, Northwest Territories, Canada.

Weber, W.A. 1987. Colorado Flora, Western Slope, University of Colorado Press.

Weber, W.A. 1990. Colorado Flora, Eastern Slope, University of Colorado Press.

Webster, J. 1980. Fungi, 2nd Edition. Cambridge University Press.

Weedon, N.F. 1996. A Sierra Nevada Flora, 4th ed. Wilderness Press, Berkeley, California.

Weiner, M.A. 1972. Earth Medicine-Earth Food: plant remedies, drugs, and natural foods of the North American Indians. MacMillan Publishing Company, New York.

Welsh, S.L., N.D. Atwood, S. Goodrich, and L.C. Higgins. 1993. A Utah flora. 2nd edition revised. Print Service, Brigham Young University, Provo, UT. 986 pp.

Wherry, E.T. 1942. Go Slow on Eating Fern Fiddleheads. American Fern Journal 32(3):108-109

Whipple, J.J. 2001. Annotated Checklist of Exotic Vascular Plants in Yellowstone National Plants. Western North American Naturalist 61(3):336B346

Whipple, J.J. 1999. The Yellowstone Sand Verbena. Castilleja 18(4):1-3

Williams, R. L. 1984. Aven Nelson of Wyoming. Colorado Assoc. Univ. Press, Boulder, CO, xii + 407 pp.

Whiting, A.F. 1939. Ethnobotany of the Hopi. Museum of Northern Arizona, Bulletin #15. Northern Arizona Society of Science and Art, Flagstaff, Arizona

Whittlesey, R. 1985. Familiar Friends: Northwest Plants. Portland, Oregon: Rose Press

Wilford, W.R., J.P. Harrington, and B. Freire-Marreco. 1916. Ethnobotany of the Tewa Indians. Smithsonian Institute, Bureau of American ethology, Bulletin #55, Government Printing Office, Washington D.C.

Willard, T. 1992. Edible and Medicinal Plants of the Rocky Mountains and Neighboring Territories. Wild Rose College of Natural Healing, Ltd., Alberta, Canada

Wyman, L.C. and S.K. Harris. 1941. Navajo Indian Medical Ethnobotany. University of New Mexico Press, Albuquerque, New Mexico.

Zigmond, M.L. 1981. Kawaiisu Ethnobotany. University of Utah Press. Salt Lake City.

Zwinger, A.H. and B.E. Willard. 1972. Land Above the Trees: A Guide to American Alpine Tundra. Harper and Row, New York.

INDEX

G

Q

R

www.ingramcontent.com/pod-product-compliance
Lightning Source LLC
Chambersburg PA
CBHW031821170526
45157CB00001B/142